THE IMMUNE MIND

Also by Dr. Monty Lyman

The Remarkable Life of the Skin:
An intimate journey across our surface

The Painful Truth: The new science of
why we hurt and how we can heal

The Immune Mind

The Hidden Dialogue Between
Your Brain and Immune System

DR. MONTY LYMAN

**DIVERSION
BOOKS**

Diversion Books
A division of Diversion Publishing Corp.
www.diversionbooks.com

For more information, email info@diversionbooks.com.

First Diversion Books Edition: September 2024
Paperback ISBN: 9781635769876
e-ISBN: 9781635769814

Printed in the United States of America
10 9 8 7 6 5 4 3 2 1

Diversion books are available at special discounts for bulk purchases
in the US by corporations, institutions, and other organizations.
For more information, please contact admin@diversionbooks.com.

For Rob, Hannah and Phin

Contents

Author's Note   ix
Prologue   xi

Part One: The Open Mind
  1 A Tale of Two Systems   3
  2 The Hole in the Wall   21
  3 The Sick Sense   35
  4 A Tale of a Supersystem   49
  5 Your Mind on Microbes   68

Part Two: Things Fall Apart
  6 Friendly Fire   93
  7 The Inflamed Mind   113
  8 Inflammatory Thoughts   139
  9 Nobody's Land   156
 10 The Cost of War   170

Part Three: Resetting Your Defence System
 11 The Anti-Inflammatory Life   183
 12 Eat   188
 13 Play   204
 14 Love   220

References   235
Acknowledgments   269
Index   271
About the Author   285

Author's Note

On the odd occasion when I have had to explain to a dinner-party guest that medical students do not have to take the Hippocratic Oath on graduation, there has always been a slight exhale of disappointment. But this ancient vow still underpins many of our modern codes of medical ethics, including the sacredness of doctor–patient confidentiality: 'What I may see or hear in the course of treatment or even outside of treatment in regard to the life of patients, which on no account one must spread abroad, I will keep to myself.'[1] In this book I have protected my patients with many layers of discretion: names, features and locations have been altered beyond recognition.

While my clinical and research interests explore how the mind, immune system and gut microbiome interact, I have no institutional or financial incentive to promote a particular test or treatment. Through detailed research and interviews with world experts and patients, I have developed opinions on this fascinating, fast-emerging field – opinions that are separate from the institutions with which I am affiliated: the National Health Service (NHS), the University of Oxford's Department of Psychiatry and Lincoln College. While my intention is that reading this book enriches your life and health, individualized medical advice should come from your health practitioner.

Prologue

*I am very poorly today and very stupid and hate everybody and
everything.*

CHARLES DARWIN, LETTER TO
CHARLES LYELL, 1 OCTOBER 1861

'This week's been a bit of a write-off,' she said. I paused for a second
or two. Had I done anything wrong? It had all been going so well.
Her case, which had started out so simply, would make me rethink
everything I had been taught about mental health.

It was my first year of psychiatry specialty training, and Emma
was one of my first cognitive behavioural therapy (CBT) patients. I
had recently been trained in the fundamentals of this form of psychological therapy, which culminated in treating a patient with
major depressive disorder (MDD), which most people call depression. There were twelve weekly sessions in total, each followed by a
supervision session with a senior clinician. Emma was a financial
advisor in her mid-forties and had suffered deep troughs of disabling depression for at least a decade. She was usually a high-flyer,
metaphorically and physically, regularly leaving her central Oxford
townhouse to jet off around the world to meet high-net-worth clients. And she played as hard as she worked, spending one half of her
time off in the London party scene and the other walking her beautiful Miniature Goldendoodle around leafy Oxfordshire. Her bouts
of depression always tended to 'come out of the blue ... like a

descending curtain ... it's like I'm looking at life through a grainy, black-and-white filter'.

When I met Emma for our first session – via video link due to the Covid-19 pandemic – she was in a bad place. Emma had become trapped in a tormenting cycle of believing that she was hopeless at her work and unlovable in her personal life. Her usual enthusiasm and drive had completely evaporated and she could not extract pleasure from anything, even her go-to Netflix comedy shows. Before each weekly session, Emma would enter her depression score into the hospital's online system, which we would then discuss and track over the course of her treatment. The score was calculated from the Patient Health Questionnaire, which assesses the following symptoms of depression: low mood, reduced ability to experience pleasure (called anhedonia), fatigue, sleep problems, appetite changes, reduced concentration, feelings of guilt, speed of functioning and thoughts of suicide. She scored reasonably highly in all of these except the latter, and wound up with a total of 18 out of 27: moderately severe.

Over the coming weeks, to my delight, that score began to drop. Much more importantly, Emma's subjective experience was improving alongside this relatively arbitrary symptom score. Indeed, she found it immensely encouraging to see her mood lift, slowly but surely. Emma found that CBT's pragmatic approach suited her – from finding positive evidence to put her negative thought patterns 'on trial', to engaging in meaningful activities that she was able to gradually take part in more often. It was wonderful to see. But I have to admit that my security and satisfaction lay primarily in the numbers, which I would receive just before each session. Perhaps the illusion of quantification helped soothe my insecurities about my foray into psychological therapy. I had been a late convert to the field of mental health. At medical school I had always imagined that I would become a specialist in an immunology-heavy field such as infectious disease (ID), dermatology or rheumatology. These specialisms dig deep into one of the most complex entities in biology: the immune system. I loved the brutal beauty of the diverse group of microscopic cells that make up the body's immune army. I was rapt by the ways in which immune cells detect and destroy pathogenic bacteria, viruses and parasites. Essentially, the immune system is a military organization fine-tuned by millennia of microbial wars. I even undertook a Master's in Immunology during my medical degree, looking at how a newly discovered immune cell (the innate lymphoid cell) responded to being exposed to little bits of bacterial or viral material.

But soon after I qualified as a junior doctor I decided to switch to psychiatry. It's an imperfect specialty with a chequered history, but seeing wise and empathic psychiatrists at work – in many cases helping to positively transform the lives of their patients – was compelling. I wanted to be able to come alongside patients and treat them as unique individuals, helping them from a psychological, biological and social point of view. My experience with Emma seemed to be validating this.

That was, however, until week seven. I logged into the system a few minutes before our session, my brain awash in dopamine in anticipation of her number dropping even lower than last week's score of 8. The screen flashed up with her graph: a steady improvement over six weeks followed by a sudden, violent uptick all the way to 16. This was close to her tally at the start of the course. 'This week's been a bit of a write-off,' she said at the start of our session. After a brief pause I asked why. 'Oh, not in that kind of way. It's just that I had Covid again . . .'

Thankfully her infection was mild and lasted only a few days, and by the end of her twelve-session course of CBT her depressive episode that we had tackled together had completely resolved. But a niggling thought would not leave me: how could a viral infection mimic moderately severe depression? I reflected on my own symptoms during a recent bout of the flu: low mood, anhedonia, increased sleep, fatigue and a complete loss of appetite. In fact, most of the symptoms one experiences during the acute stages of an infection seem more 'mental' than physical. Whether you have Covid, cholera or the common cold, 'sickness behaviour' – how we feel and act when fighting an infection – is universal. It seems that your immune system is recruiting your brain to produce behaviours that aid your defence: primarily, wrapping up warm and staying in bed. Indeed, an increasingly accepted hypothesis is that sickness behaviour provides a survival advantage both for the individual – in saving energy to fight off the infection – and also for the herd at large: a state of mind that makes you feel low, withdrawn and tired is one that will restrain you from heading out to the party and sharing your virus with your tribe of revellers.

But there was something else in Emma's story that stayed with me. It was a brief, throwaway comment she had made in our first session, when I was exploring her previous relapses of depression: 'It's a bit weird, but whenever my depression comes back, or whenever I'm stressed, my eczema flares up.' She rolled up the sleeve of her jumper to reveal a raw, red forearm. It was here, in a fairly unremarkable case study of someone undergoing CBT for depression, that I had my first glimpse of a 'bidirectional' relationship between the immune system and the mind. Emma's immune system had recruited her brain – more specifically her thinking, feeling *mind* – to help deal with a viral infection, but her mind had also activated her immune system, causing skin inflammation. I was peering behind the curtain of my medical training: a relationship between the immune system and mental health was completely absent from all of my medical school textbooks.

In this book we will explore the exciting new science revealing the strength of the immune-mind connection, what happens when it goes wrong and how we can use this knowledge to improve our mental and physical health. Aside from sickness behaviour, which most of us have experienced at some point, doctors have long observed other ways in which the immune system can alter the mind, from the hallucinations and acute confusion of delirium common in the frail and elderly, to the varied (and often extreme) psychiatric manifestations of untreated HIV and syphilis. But, until very recently, it has been almost heretical within the medical profession to hold the view that our immune system can cause or worsen mental disorders. Even today, when I tell other medical professionals that my clinical and research interests cover the relationship between mind and immune system, I often get a look of half-pity, half-puzzlement, as though I had planned on specializing in palm reading. Resistance to the immune-mind connection is, understandably, stronger within the old school. A week before writing this, a retired neurologist advised me at a dinner to 'file that under "interesting but implausible" . . . you'll be wasting your time, boy!' As we will see, the unwillingness to explore connections between the mind and the body stems from a way of thinking called 'mind-body dualism': the body being a physical machine, whereas the mind is of a completely separate essence. We tend to treat organs and body systems as though they are separate components of a vehicle – finely crafted for a specific purpose but assembled right at the last minute: the stomach a food macerator, the heart a blood pump. The idea that two systems might work in tandem for a particular cause, or might influence each other, is complex and cognitively uncomfortable. The truth of the matter is working against tradition, against a historical absence of the required scientific techniques and against the powerful human need for simplicity and categorization.

In this book I want to rip open the curtain we have erected between the mind and the body. By 'we', I am mainly taking aim at the post-enlightenment West and the medical profession. I do this firstly because I myself am embedded in both of these cultures. But I also believe that the many wondrous developments in Western medical science over the past two centuries have come at the cost of putting the messiness of human psychology to one side. I believe that we, as a society, completely misunderstand the relationship between mind and body. The implications are not just academic: I believe that misguided beliefs are causing people preventable and relievable harm. More and more people are experiencing – and more and more doctors are witnessing – conditions that cannot be neatly associated with specific organs, symptoms or specialties. Individuals living with long-term conditions, and those treating them, too easily fall into the trap of labelling something as 'physical' or 'mental'. With the former, we can underestimate the power of the mind to change our body and our experience of it. With the latter, we can fail to realize how our mind is often at the whim of unseen bodily processes. The reality is that there is no mental disorder that is not also physical, and most physical diseases have some mental element to them. We have been trained to pigeonhole disease into either one or the other, even to the extent that we visit one hospital for the body and another for the mind. I know from my experience as a doctor that there are both implicit and explicit pressures to force patients down the grooves of either 'physical' or 'mental', even when it is clear that neither is a perfect fit.

I originally set out to write this book about two medical fields: immunology and psychiatry. I wanted to explore how they developed and grew in different directions, and how exploring the new science behind the immune-mind connection may help our understanding of mental health. But instead I've ended up discovering so much more. It is my contention that your mind and your immune system are not simply linked, but can be viewed as part of the same system: I will call this the 'defence system'. This system functions to protect us from the horrors of the outside world, whether they are bear-sized or bacterium-sized. The immune system and brain work together to differentiate friend from foe, and to mount an appropriate response to the latter. I believe that viewing these seemingly separate systems as a unified 'supersystem', so to speak, is a clear and comprehensible way of escaping the mind-body dichotomy and cultivating a more accurate view of how the human organism works. This explains why our immune system so often influences our thoughts and feelings, as well as why psychological stress and trauma affects our body's immune army. Not only does knowledge of this defence system revolutionize our understanding of health, but it should also change how we view and treat disease.

Join me in ripping up the curtain that society has placed between the mind and the body. In Part 1, 'The Open Mind', we will explore the beauty of the mind-immune connection, and how humankind's vital diplomatic relationship with the microbial world has moulded our minds in unexpected ways. It will also aid us in conceptualizing the defence system, which will help us answer questions ignored or unanswered by medical orthodoxy: Why do our moods and behaviours change when we're sick? Why are so many of our current psychiatric treatments ineffective? Why do purely psychological stressors activate our immune system? And how does mindset play a role in reducing the risk of disease and in recovery? In this first part of the book I draw from incredible discoveries from the past decade that are demonstrating how the mind and immune systems are intricately linked. We will also see that the human mind is moulded by the immune system's primary targets: 'microbes', also known as 'microorganisms'. These are, simply, living beings too small to be seen by the naked eye. We are only just beginning to understand how these microscopic organisms – including bacteria, viruses and parasites – affect mental health. Perhaps the importance of microbes in human health is not too surprising when we consider their ubiquity: microbes were around long before humans, and we harbour more microbial cells in and on us than we have human cells: a vast community we call the 'microbiome'.

The book's second section, 'Things Fall Apart', builds on this new knowledge of mind and body to see what happens when our defence system goes wrong: from depression to dementia, the effects of psychological stress on the body to the symptoms of post-viral illness.

We will see that the modern world is unrecognizably different from the one in which our defence systems developed, and the increases in certain mental disorders, allergies and autoimmune disorders are all the result of unbalanced defence systems.

Finally, in 'Resetting Your Defence System', we look at practical ways in which we can improve our mental and physical health by rebalancing a miscalibrated defence system.

I hope that you find this book to be a journey, from seeing yourself as a mind trapped within a machine-like body to viewing both mind and body as utterly intertwined. I hope that through the history, anatomy, philosophy and physiology you encounter in the following pages, you cultivate a grander view of life and uncover a deeper love of what it means to be human. We need an open mind, because the mind itself is open.

PART ONE

The Open Mind

1

A Tale of Two Systems

*How neurology and immunology
grew up, and grew apart*

> *Ships that pass in the night, and speak each other in passing,*
> *Only a signal shown and a distant voice in the darkness;*
> *So on the ocean of life we pass and speak one another,*
> *Only a look and a voice, then darkness again and a silence.*

HENRY WADSWORTH LONGFELLOW,
A THEOLOGIAN'S TALE

The nervous system

A walk down Oxford's High Street is a sensory banquet. As you follow its gentle curve, it feels as though you are gradually unfurling an ancient scroll. Jagged gothic pinnacles break up the skyline. Well-weathered statues, staring vacantly, stand aloof in their niches. Rickety, timber-framed houses rub shoulders with imposing stone colleges, bunched up along the street like awkward professors gathered for a group photo. But any illusion of being transported back in time is frequently shattered. Open-top tour buses huff as they pick up their passengers. The pavements writhe with lost tourists and tipsy students diffusing along the city's veins. Amid this bustle, even a local who's walking the length of 'The High' could be forgiven for missing many of the narrow medieval alleyways that slip off the sides of the busy street. One of these passages leads down to an old, five-gabled townhouse that – in one of its various guises across the centuries – is now the excellent Chiang Mai Kitchen restaurant. Visitors can be certain of delicious Thai cuisine, but there is something of which they may not be aware. As they cut into their massaman chicken or de-vein the prawns in their tom yam soup, today's diners are maintaining the building's long tradition of dissection.

In the mid-1600s, this hall was the workplace of Dr Thomas Willis and his recently deceased clients.[1] This intrepid medic would spend most of his clinical days out in the market squares of Oxford and its neighbouring towns, competing with other physicians for the attention of the sick. But Willis would also keep an eye out for local executions; he would hang around waiting for fresh bodies to bring back to his workshop. There he would gently prise open their skulls and explore the mysterious, mushy structures that lay within, in an attempt to map the largely unknown territory of the brain. Willis's efforts were aided by a remarkable set of friends. Robert Boyle suggested soaking the brain in port, its high alcohol content preserving the samples.[2] (Boyle – now most famous for his eponymous law that measures the relationship between the pressure and volume of a gas – went on to become a founder of modern chemistry.) Once the specimens had been preserved, Willis had another friend, Christopher Wren, draw them in unprecedented detail, foreshadowing his development into one of the most celebrated and prolific architects in history.

Willis was a pioneering neuroanatomist, describing and naming numerous brain structures and identifying all twelve cranial nerves in incredible detail, laying the foundations for modern medical textbooks.[3] He also coined the term *neurologia*, from which we get *neurology*. But, perhaps most importantly, he was also an excellent clinician, and connected his findings in the brains of the deceased with the conditions he himself had witnessed in their lives. Linking brain and behaviour, Thomas Willis identified the cerebral cortex – the grey matter that forms the thick outer layer of the brain – as 'the primary seat of the rational soul in man ... and the source of movements and ideas'. He based this on differences in cortex size and shape in brains of those who had lived with learning difficulties.[4] This was revolutionary, considering that the contemporary consensus held that the cortex was an intellectually unimportant collection of blood vessels.

He also located the mind within the brain, which was not a given – in the seventeenth century, the centuries-old search for the source of human thought was still the topic of intense debate. In nearby Stratford, a certain William Shakespeare had recently summarized it well: 'Tell me, where is fancy bred [. . .] in the heart or in the head?'[5] Both Team Heart and Team Head had a venerable Greek philosopher-scientist to champion their cause. Aristotle, who had been right about so many things, placed the mind in the heart, and relegated the brain to a cooling system. Galen, on the other hand, placed it in the fluid-filled ventricles of the brain. While Galen was much closer, neither was correct. When Thomas Willis enrolled in medical school in the 1640s, he was committing to roughly fourteen years of rote-learning the works of Aristotle and Galen. His studies were cut drastically short, however, when the English Civil War broke out. His inquisitive mind, therefore, had to be nurtured not by pure reason, nor by regurgitating the classical canon, but instead by putting his eyes to work by observing real people and getting his hands dirty with real human bodies. Perhaps he was saved from medical school.

Many of Thomas Willis's incredible discoveries were published in his *Cerebri Anatome* in 1664. I headed to the Special Collections archive of Oxford's St John's College to have a look at a first edition. As I delicately leafed through the pages, under the justifiably watchful eye of the librarian, I was amazed at how little modern textbooks have added to the detail, and how much they have taken away from the artistry. Wren's illustration of the underside of the brain is drawn with the same attention and reverence he would later pay to cathedral vaults. *Cerebri Anatome* was a groundbreaking publication, and it could be argued that 1664 marked the victory of Team Head – the source of human thought, emotion and movement was now firmly within the skull.

Willis's anatomical and clinical findings would help pave the way for further astounding discoveries in neuroscience, too. In the early nineteenth century, British anatomist Charles Bell and French physiologist François Magendie independently found that specific nerves conveyed sensory input to the spinal cord, while others transmitted motor (movement) impulses. Towards the end of the nineteenth century, advances in microscopy and cell staining enabled the Spanish neuroscientist Santiago Ramón y Cajal to identify the individual component parts from which the brain and the rest of the nervous system are built: neurons. Neurons (also called nerve cells) are cells that typically consist of three parts: the cell body (containing the nerve's DNA and most of its cellular machinery), the dendrite (the branch that leads from the periphery into the cell body) and the axon (the branch that travels away from the cell body). Ramón y Cajal's beautiful drawings of the treelike projections of nerve cells are among the most celebrated illustrations in the history of science. By the turn of the twentieth century it was known that neurons are electrically excitable cells that make up the brain, spinal cord and peripheral nerves, and that they relay electrical signals to each other. Another crucial discovery was that the axon of one neuron does not touch the dendrite of the next; the neurons are separated by a tiny, microscopic space called a synapse. When the electrical impulse reaches the end of the first nerve, it causes the release of chemicals called neurotransmitters. These travel across the synapse, which activates the next nerve, which in turn activates the next, and so on. This anatomical and physiological framework provides the basis for modern neuroscience.

As we assemble the elements that make up a functioning brain, we find that we're breathing life into our bodies, too. Neurons join together to form an intricate web of electrically active tissue – a sort of circuit board, collectively called the nervous system. This can be divided into the central nervous system (CNS), which consists of the brain and the spinal cord, and the nerves outside of these structures, the peripheral nervous system (PNS), innervating the whole human body. The PNS is itself divided into two main parts: the somatic nervous system, which deals with delivering sensory information to the brain and controlling voluntary movements, and the autonomic nervous system, which controls largely unconscious bodily functions, from breathing to the fight-or-flight response.

To show why the nervous system matters, let's give you a paper cut. Suppose that, as you turn this page, you let your mind wander and absent-mindedly allow the tip of your index finger to slide along its edge. With the right amount of pressure, at just the right angle, the paper becomes a razor blade, which slices into the nerve-rich skin of your fingertip. Before you experience any pain – and even before you know what's going on – your arm rapidly retracts. Next, you feel a sharp, throbbing pain in your finger. So, what's going on? Danger receptors embedded in your skin (called nociceptors) are activated and send an electrochemical signal across the surface of a neuron, to which the receptors are attached, called a sensory nerve. Travelling at a respectable pace of around 20 metres per second, the signal quickly reaches the spinal cord. Here it encounters synapses, across which it passes the signal to nerves going in different directions. Some of the electrical impulses travel from the sensory nerve to a motor nerve, which activates muscles in your arm, causing your hand to withdraw from the offending object before any nerve impulses reach the brain. This is a reflex: automatic, unthinking and unknowing. Other nerve impulses travel up the spinal cord, relayed from neuron to neuron, into the brain. The brain integrates all of this sensory information – what happened and to which part of the body – with other contextual clues, from your current mood to your memories of similar past experiences. The brain then decides, outside of your conscious awareness, that a breach in your skin's defences is a threat to the body, and thus sets off an alarm that we experience as pain. This miracle of computing occurs within milliseconds. Just a couple of seconds after cutting your finger you feel foul, frustrated at your careless motor coordination and inattentiveness. You then begin a learning process, promising yourself to be slightly more careful next time you turn a page. The experience consciously and subconsciously changes your behaviour.

This is the grossly simplified version of the main functions of the nervous system outlined in textbooks: it receives inputs from the body and the environment, integrates these in a meaningful way, and then acts on them. It detects whether a sensory input poses a threat to the body or not, and then it acts accordingly. The nervous system is conventionally seen as a highly complex system – with the brain as its command centre – that coordinates our sensations, controls our movements, and animates our thoughts and feelings. In sum: the brain decides, and the body obeys.

The immune system

On leaving Willis's cadaveric curry house and re-entering The High, a two-minute stroll eastwards takes you to a grey plaque, flush against a wall of honeyed Cotswold stone. Engraved in golden script, it reads:

In a house on this site between 1655 and 1668 lived

ROBERT HOOKE

*Inventor, scientist and architect, who made a
microscope and thereby first identified
the living cell.*

In 1653, Hooke took up a chorister's place at Oxford and soon became employed by Thomas Willis as his chemical assistant. Here he would be introduced to Willis's circle of unfairly talented polymaths, including Robert Boyle and Christopher Wren. This intellectually invigorating environment nurtured a mind that would make several scientific breakthroughs in a number of fields, from physics to palaeontology. But Hooke is best known for his use of the light microscope, an instrument newly invented in the Netherlands. In January 1665 – within a year of *Cerebri Anatome*, Willis's medical masterpiece – Hooke published *Micrographia*. In this book, which became the first science bestseller, Hooke describes the wonders of a previously undiscovered microbial world. He was a gifted artist and his detailed drawings of the microscopic – from silkworms to a single layer of cork – can be admired for free in the Royal Society's online archives.[6] These pictures are matched by entertaining prose: he likens the humble flea to a knight 'adorned with a curiously polished suit of sable armour, neatly jointed, and beset with multitudes of sharp pins'.[7] *Micrographia* opened up an invisible world to the public; Samuel Pepys described it as 'the most ingenious book that I ever read in my life'.[8] But the bestseller was also a landmark scientific publication. It introduced the term 'cell', the building block of all life forms. Hooke is also the first person to describe a specific microbe: the *Mucor* fungus. A decade after the publication of *Micrographia*, the equally innovative Antonie van Leeuwenhoek – a self-taught scientist from the Netherlands – would observe and describe mobile microorganisms, which he called 'little animals'. These included the first bacteria ever to be seen by the human eye. While most scientists during the Scientific Revolution looked up at the celestial bodies, these pioneers instead looked down to wonder at an abundant microbial universe.

These discoveries laid the foundation for microbiology – the study of microbial life – and, ultimately, immunology – the study of our bodies' interaction with these microbes. But the discovery that microbes can cause disease was two centuries away. Perhaps Hooke would not have described his fleas in such endearing terms if he had known they were the vector for *Yersinia pestis*. This bacterium was responsible for the Great Plague of London, which would wipe out a quarter of the capital's population in the year following the publication of *Micrographia*.[9] The prevailing hypothesis behind spreadable disease at the time – and one that seems to have dominated in most cultures throughout history – was 'miasma theory'. This held that disease was spread by bad, polluted air – the word 'miasma' comes from the Greek for 'pollution'. There are even fossils of this theory left in language: 'malaria' simply means 'bad air'. While we now know this to be completely wrong, miasma theory was a reasonable inference in a world with no evidence of, and no reason to believe in, microscopic organisms.

There had long been clues in Greek, Hebrew and early Islamic texts that humans could directly spread disease to each other,[10] but even by Hooke's time, 'contagion theory' had never really caught on. However, probably around the time Hooke identified the first cell, another revolution in the history of infection and immunity was taking place. This revolution did not emerge in a European metropolis, but instead a village in India or China that has long been lost to history. Someone had realized that smallpox could be prevented.[11] The method? Scraping the skin, or collecting the pus, of smallpox scabs on an infected individual, then respectively either rubbing or smearing this into cuts made in the skin of the uninfected. This process, called variolation, proved remarkably effective at preventing one of humanity's deadliest diseases; infection before variolation came with a 30 per cent chance of death. This discovery came about with no knowledge of the existence of the smallpox virus and no knowledge of the immune system that brought about this treatment's success.

Variolation's first scientific test came in 1721. Lady Mary Wortley Montagu, wife of the British ambassador to the Ottoman court in Constantinople, returned from the East as a fervent advocate of this mysterious treatment. Lady Montagu had watched her brother die of smallpox and, although surviving it herself, had been left with a face disfigured by pockmarks. In the Ottoman Empire, she spent extensive time with local women to explore and record their customs, and came upon the practice of variolation. Eager for her children to be spared her fate, she had the embassy's doctor, Charles Maitland, inoculate her son in Constantinople, followed by her daughter when they travelled back to Britain. She next convinced the British royal family to test the safety of variolation, coaxing the Princess of Wales to take part in one of the first recorded clinical trials. Six prisoners on death row were made a relatively attractive offer: if they agreed to have pus from a smallpox patient rubbed into an incision made in their skin, those surviving would avoid the noose. All fell sick for a couple of days, but all recovered, and all were released. The Prince and Princess of Wales would soon have their own children inoculated against smallpox, and variolation took off.

Variolation was not without its dangers, however, with a small but significant number of those inoculated dying, including one of George III's sons. It took another fiercely observant pioneer to find a safer, more reliable solution. At the end of the eighteenth century, Dr Edward Jenner was doing his usual medical rounds in the deep Gloucestershire countryside when he realized he had been looking at the solution every day. As he strolled down the country lanes, Jenner realized that milkmaids were the only people whose skin was not scarred by smallpox. He hypothesized that they contracted cowpox, the milder bovine equivalent of the disease, which created immunity to smallpox.[12] Testing his theory, he took pus from Sarah Nelmes, a milkmaid with cowpox, and injected it into a cut in the arm of a local village boy, James Phipps, who had not yet contracted smallpox. A few days later, Jenner injected the boy with scab material of someone infected with smallpox. Nothing happened. This eureka moment led to the world's first vaccine (*vaccus* being the Latin for cow).[13] Jenner has arguably saved more lives than any other scientist in history.

While vaccination was shown to work even in the absence of any knowledge of *how* it worked, we would not stay in the dark for long. The nineteenth century saw the death of the miasma theory of disease and its replacement by the currently accepted hypothesis: 'germ theory'. In its simplest terms, this is the hypothesis that most spreadable diseases are caused by infection with particular microbes. In 1876, the German doctor and microbiologist Robert Koch discovered that anthrax was caused by a specific bacterium, *Bacillus anthracis*. He would go on to do the same with other infectious giants: tuberculosis (*Mycobacterium tuberculosis*) and cholera (*Vibrio cholera*). Koch, along with his rival and equal Louis Pasteur, completely changed the medical landscape. It became clear that infectious illnesses were the result of disease-causing microorganisms, which we now call pathogens. Hot on the heels of the germ theory revolution, the turn of the twentieth century saw the discovery of how our body responds to these microbes: the immune system. The significance of this achievement is marked by the awarding of the 1908 Nobel Prize to two remarkable scientists. Élie Metchnikoff, a Ukrainian scientist who would go on to work under Louis Pasteur, observed bacteria being engulfed and destroyed by unusual white blood cells we now call macrophages (big eaters). At the same time, Paul Ehrlich was working with his compatriot Robert Koch, going on to successfully treat diphtheria not with immune cells, but with blood plasma full of large proteins produced by immune cells. He would term these 'antibodies'.

Over the course of the twentieth century, curious minds and technological advances began to explore the vast, hidden world of the immune system. And it was beautiful. As each interaction of the immune system was painstakingly revealed, it became apparent that this is one of the most complex networks in nature. Your immune system is an army made up of billions of diverse cells with an exquisite skill for distinguishing between 'self' and 'non-self', whether non-self be bacteria, parasites, viruses, cancer cells or splinters. These cells communicate with each other – often across considerable distances – to initiate and then rein back complex defensive responses on an orchestral scale. The immune system blends the balance of a ballerina with the offensive capabilities of an elite Marine unit. Once intruders are found, they are destroyed with brutal diversity: some immune cells eat pathogens; others trap them in nets of DNA; others secrete molecules that bring about the programmed death of infected cells. Some proteins of the immune system latch on to viruses to stop them entering cells; others can form 'attack complexes' that punch holes in the sides of bacteria, leaving them to fill up with water and explode. One might almost begin to feel sorry for the pathogens.

When Robert Hooke took his first glimpses into the microscopic world in the seventeenth century, he could not have foreseen its sheer immensity. Unknown to him, inside his own body wars were being waged; pathogens and immune cells engaged in an evolutionary arms race. By the early twentieth century we finally had an explanation both for how we catch infections and fight them off. It also became clear how vaccines work: if our body is exposed to an inactivated piece of a bacteria or virus, our immune system identifies it as foreign and kicks into gear, without the body becoming infected. The immune system triggers an initial inflammatory response and, in the long run, it remembers the specific biological patterns and structures of the pathogen. This enables the immune system to rapidly eliminate the microbe if it enters our body again. It also became evident that the immune system can expertly detect pathogens that have never been seen by the body before. By the 1980s, the case was closed: we knew, in general terms, why we have an immune system and how it works. But one scientist found that one of the key pieces of the jigsaw just would not fit.

Charles Janeway, a professor of immunobiology at Yale, was frustrated that we still could not explain why vaccines worked. In the early twentieth century, vaccinologists wanted to move on from exposing people to Edward Jenner's messy cow pus and instead inject only a purified protein from the specific pathogen of interest. But, remarkably, these targeted vaccines produced little to no response. After a bit of trial and error, it appeared that the vaccines that worked contained not only the necessary bit of pathogen, but also a generic 'messy' substance that had been used in the purification process – ranging from aluminium salts to paraffin oil. But when Janeway asked why these substances were necessary for an immune response, no one could provide a satisfactory answer. Janeway argued that, as well as having receptors that distinguish self from non-self, there must also be receptors that detect threat.[14] For example, food is clearly non-self, but it isn't usually a threat; we don't mount a full-on inflammatory response every time we bite into a burrito. But the messy substances in vaccines were a threat that our body could recognize and respond to. Janeway called these hypothetical threat detectors pattern-recognition receptors (PRRs) and argued that there must be a range of receptors that come ready-made from our genetic code, evolved to recognize generic molecules and motifs common to most pathogens, termed 'molecular patterns'. PRRs are essentially preconfigured barcode-readers that recognize specific danger patterns and then initiate an immune response. This then enables other elements of the immune system to identify the specific pathogen, respond to it and, eventually, remember it. Janeway's theory would soon be proved right: in 1998 a PRR was identified that recognized a molecule common to the outer membrane of many types of bacteria.[15] The spark that ignites the immune response had been found.

Today, the immune system is usually conceptually cleaved in two: the 'innate immune system' and the 'adaptive immune system'. The innate immune system responds to pathogens in an immediate, non-specific manner, initiated by immune cells at the site of infection. The adaptive immune system is slower but more specific, consisting of cells that can remember a specific pathogen over a long period of time. To visualize this, and to establish a basic understanding of how the immune system works, let's go back to your paper cut.

As you absent-mindedly drag the tip of your index finger down

the side of the page, the razor-sharp
of skin. As you do this, a numbe
dislodged from the page and
Most are harmless, but among th
it lands in the squishy bed of con
layers of your skin, it is quickly detected
include macrophages, the immune cells h
nikoff.* One of the pattern-recognition recep
macrophage locks on to a common molecule fou
rial cell walls. This is the threat detection that C
predicted. The innate immune system has now been tri
mêlée ensues. The macrophages promptly secrete tiny so
teins called cytokines. This broad group of molecules is abs
crucial to the immune system: they can travel both locally and
over the body, interacting with other immune cells. The immune
system is a gossipy community, and one of its main languages is
cytokine. In this case, the activated macrophages spurt out pro-
inflammatory cytokines: molecules that trigger an inflammatory
response. These recruit other immune cells (the most well-known
being neutrophils) to the site of injury. These newly recruited cells
will be responsible for eliminating any other rogue pathogens that
enter the cut, but they are also a disaster-response team, initiating
the wound-healing process. Other local immune cells get in on the
action, too. Mast cells are the spherical, spotted landmines of the
immune system. When activated, they pour out a potent cocktail of
enzymes, including histamine. These enzymes are responsible for
the red wheals of an allergic reaction, but in a response to an infection their main job is to dilate the local blood vessels, widening the
roads for incoming battalions of immune cells. The overall result is

---

* Skin-resident macrophages are known as Langerhans cells, which were
identified in the 1860s by the twenty-one-year-old German medical
student Paul Langerhans. The definition of a *wunderkind*.

the body's complex, powerful response to infection
inflammation and its effects can take many forms,
on where it is in the body and how long it persists. In the
ee many of what are termed the 'classical' features of
tion: swelling, redness, heat and pain.

e all this is going on, the macrophage begins to do something
astounding. It takes on a dual role as soldier and war reporter.
t engulfs the bacterium and uses its stores of powerful enzymes
elt the pathogen down into a mélange of constituent proteins.
me of these tiny bacterial fragments – called antigens – are unique
this particular species of bacterium. In the case of your paper cut,
the macrophage then begins a journey from the wound site on the tip
of your index finger, travelling along your arm through 'lymphatic
vessels', the highways of the immune system that connect the lymph
nodes, small bean-shaped (and roughly bean-sized) organs dotted
throughout the body. These unassuming lumps are the military gar-
risons for the immune system's most specialized combatants. The
macrophage – having lain down its weapons and now acting as war
reporter – enters these forts, holding up the bacterial protein on its
surface like a piece of the enemy's uniform. Here the macrophage
begins to interact with cells of the other, specialized half of the
immune system: the adaptive immune system. To continue the military metaphor, this recruits cells known as T-cells, which are the
immune system's Special Forces soldiers. Each T-cell has a unique
receptor for detecting potential pathogens, and the macrophage needs
to find the specific T-cell that matches its captured bacterial protein.
Different T-cells begin to pass by the macrophage, taking hold of the
fragments of bacterial uniform in the war reporter's hands, assessing
whether they are suited to fight that specific foe. Passing by the macrophages, the T-cells decide whether their receptor is the correct fit
for the bacterial protein on the macrophages' surface. Eventually –
usually after a few hours – T-cells will arrive that have receptors
exactly the right shape, and the cells bind together. Through a series

of astoundingly complex interactions, many of which we do not yet fully understand, the macrophages present a 'snapshot' of the battle – showing where on the skin it is taking place and which enemy is involved – to these T-cells. The T-cells are then able to signal to other cells and organize a coordinated immune response against any invader.[16] This process also engages B-cells, which produce antibodies directed against the specific bacterium detected breaching the skin. Antibodies are immune proteins that are often mistakenly described as bullets or heat-seeking missiles that destroy pathogens. In reality, they are more like toy arrows, with a sucker designed to attach to a specific molecule. Once stuck to its target, a B-cell can either block its receptor site (effectively rendering it useless) or flag it for destruction by slower but deadlier immune cells. An even more remarkable feature of this response is that many T- and B-cells develop a 'memory' of the specific species of bacterium in question, so that if it breaches the body's outer defences in the future, it can be dealt with swiftly.

Only the brain rivals the immune system when it comes to sheer complexity. What is rather striking, though, is how similar these two systems are. They both work to differentiate self from non-self, and also to sift out threatening cues in the environment from non-threatening ones. In essence, whether dealing with the massive or the microscopic, their job is to recognize, react and remember. As you get your paper cut, both systems kick into action. One might assume that two systems so alike and with shared goals – defence and survival – might communicate with each other. But, only until very recently, conventional thought was that the brain and immune system lived parallel lives. To understand how we ended up in this predicament, we need to go back to the seventeenth century.

The divided human

Let's jump back to 1664. Thomas Willis has published his *Cerebri Anatome* and in a few months' time Robert Hooke's *Micrographia* will arrive on the scene. But, across the English Channel, another book by another prodigious polymath has just been published: the philosopher René Descartes' *L'Homme*, or *Treatise on Man*. This book, collated and printed fourteen years after Descartes' death, is a largely philosophical case for placing the mind within the brain. Descartes believed the seat of the human soul lay in the pineal gland, a tiny pine-cone-like structure (*pinea* is the Latin for pine cone) that lies right in the centre of the brain. Locating the mind in the brain was a significant achievement – and we can forgive Descartes for not knowing that the pineal gland is actually a sleep cycle regulator. However, Descartes began to expound on an error that still haunts medicine today: mind-body dualism. He argued that the pineal gland was the nexus by which the mind – which he believed to be an ethereal, disembodied substance – interacted with the physical body. The mind and body, he claimed, were separated by both location and essence: a material, machine-like body is controlled by an immaterial mind. This proposal, known as Cartesian dualism, produced more questions than answers – namely, about how these two separate entities interact. There were, however, benefits to viewing the body as a machine. In leaving the mind to one side, Descartes had made a Faustian pact that would enable scientists and physicians to study the body's organs without worrying about the chaos and complexity of the human mind. But, although it had its uses, Descartes' theory of mind-body dualism created a conceptual cage that would trap medicine in erroneous thinking for centuries. The idea that the mind and body are completely separate entities pervades modern medicine – particularly Western medicine – to this day. We have our 'physical health' conditions tended to by technician-like physicians and surgeons, our 'mental health' conditions by psychologists and psychiatrists. We go to one hospital for the body, another for the mind.

The privileged brain

'. . . And, of course, we don't have to worry about the brain, as it is "immune privileged". . .'

It was one of my first medical school lectures, and a tweed-clad, round-spectacled immunologist was taking us on a roving tour of the immune system. It seemed to pervade every organ and orifice of the human body. That is, except the brain. Immune privilege, he explained, meant that the brain's tissue is too precious to allow it to be affected by the collateral damage of inflammation, so it thus neither has resident immune cells, nor does it allow immune cells into its tissues.

This wasn't just conjecture; it was based on various lines of evidence built in the twentieth century. It was known that the immune system treats cancer cells as non-self, just like pathogens, so it plays a key role in destroying tumours in the body. In the 1920s, Japanese scientists found that if you put tumour cells into a rat's body, immune cells promptly launched an attack on these cells, but if the tumour was placed in a rat's brain, it appeared to survive.[17] It seemed that the immune system couldn't reach inside the brain. There were also two clear anatomical explanations for the brain's unique, isolated position. The first is that the brain didn't appear to have lymphatic vessels, which suggested that it wasn't hooked up to the immune system's network. The other is the 'blood-brain barrier'. This is a thick, multicellular wall between our circulating blood and our brain. Some very small chemicals can passively diffuse across it – alcohol and caffeine need no introduction. But otherwise it was considered impermeable to larger molecules and cells, such as those of the immune system. It was understandably assumed that immune activity in the brain occurred only in the context of severe disease, as the result of a catastrophic breakdown of the blood-brain barrier.

This 'Berlin Wall' between the brain and the immune system is conceptual as well as structural.[18] While the nervous system (the brain, spine and nerves) and the immune system play similar roles, it has always been assumed that they are separate systems designed to defend against completely different threats. The understanding has been that the brain uses senses to recognize and react to 'macro' threats – the sound of a lion's roar, the sight of a rival tribesman running at you with a spear, the sensation of searing heat as your hand lingers on the saucepan handle a little too long; the immune system, on the other hand, is on the lookout for 'micro' threats. The two systems have been treated like the army and the navy: a shared overarching purpose – defence – but achieved in very different ways and in completely different theatres. In sum: the systems are similar but separate – indeed, why should our brain and immune system even need to work together? Why should our mind and our mental health be influenced by our immune defences, or vice versa? These assumptions have had a long, strong grip on the medical profession. At medical school, my neuroscience and psychology textbooks spilt no ink on the immune system. Equally, my treasured *Janeway's Immunobiology* contained no mention of the brain or the mind.

As Thomas Willis and Robert Hooke sat down at the end of the day to share findings and propose theories – perhaps aided by the remainder of Robert Boyle's brain-staining port – I doubt they had a full grasp of the magnitude of their discoveries. Willis had carefully tended the vine of neuroscience, from which neurology, psychiatry and psychology would fruit. Hooke had laid a microscopic seed that would grow into the branches of microbiology and immunology. As these branches grew, however, they followed the trellis of mind-body dualism built by Descartes; the study of the immune system and the mind grew in different directions. For centuries, it was thought that these systems had no reason to be connected. But, as we are about to see, we are just starting to discover that these systems are much more intertwined than we could ever have imagined. We will start with something utterly astounding: a recent discovery of new anatomy, a literal bridge between the brain and the immune system. This promises a complete revolution in our understanding of mind and body.

2

The Hole in the Wall

How hidden anatomy and secret physiology are changing medicine

> *There is nothing like looking, if you want to find something. You certainly usually find something, if you look, but it is not always quite the something you were after.*
>
> J. R. R. TOLKIEN,
> *THE HOBBIT*

It couldn't be true. That shouldn't be there.

It's the summer of 2014. On a muggy evening in Charlottesville, Virginia, a lab researcher in his late twenties is rubbing his eyes. Could what he just saw under the microscope really be new anatomy? Whether he knew it at the time or not, this observation would change the way we see our brains, our bodies and – ultimately – ourselves.

The 1980s and 1990s had already witnessed huge leaps in our understanding of individual components of the immune system, as well as remarkable advances in neuroscience and brain imaging. But any suggestion that these systems should have a working relationship was flirting with heresy. However, some, such as Michal Schwartz from the Weizmann Institute in Israel, dared to question the canon. In 1999 she and her colleagues found that T-cells helped to protect crushed neurons in the central nervous system from further degeneration.[1] It was also observed that mice with impaired immune systems developed neurodegenerative diseases (such as motor neuron disease) much faster than normal mice did.[2] But while these findings sent ripples through the neuroscience community, the concept of a completely immune-privileged brain, hiding behind the impenetrable wall of the blood-brain barrier, continued to reign supreme.

I didn't question the orthodoxy when I was at medical school, although there was one thing that didn't sit right with me. In peripheral injury and inflammation – let's think back to your paper cut – I was taught that nerve endings can release inflammatory substances, such as 'substance P'.[3] This molecule can interact with immune cells, and, among other things, cause mast cells (immune-system 'landmines') to burst and ooze out their potent inflammatory potions. Conversely, it seemed that cytokines produced from immune cells could interact with nerve endings, causing them to increase pain sensitivity in the injured area. To me it seemed that, in the peripheries at least, the immune system and the nervous system spoke each other's languages. Sometimes I wondered whether similar neuro-immune conversations were going on inside the brain, conversations that worked towards a shared goal of protecting the body from infection and injury. Nonetheless, I would trust in my textbook, which spoke of the brain as an all-ruling command centre, sheltered from the microbial wars being waged within the body.

But things look very different now. The last decade has witnessed a whole revolution in our understanding of how the brain and the immune system communicate – and I do not use the 'R-word' lightly; this is no hyperbole. The revolution has been driven by determined, observant individuals and aided by dramatic improvements in research technology. It has revealed new anatomy, new cells and new pathways inside and outside the brain. Let's begin our journey just outside the brain, in its borderlands: a strange, unruly place where few have looked before.

On the brain

Between the brain and the skull lie the meninges. These are fibrous layers that wrap around the brain, protecting and cushioning the delicate and valuable brain tissue within. This protective function has given the name 'mater' (mother) to the three meningeal layers. The innermost layer is the 'pia mater' (Latin for 'tender mother'), a thin, delicate, translucent layer that sticks to the outer surface of the brain like cling film. On top of the pia mater is the 'arachnoid mater', which is more like a thin, loosely fitting plastic bag. Between the arachnoid mater and the pia mater is an all-important space, the 'subarachnoid space', which is filled with a clear liquid called cerebro-spinal fluid, or CSF. The arachnoid mater is bound to the pia mater by web-like structures that cross this space, resulting in the Latin for this layer, literally meaning 'spidery mother'. These two thin layers are very similar, both deriving from the same embryonic neural tissue as the brain – some consider them to be a single structure. The outermost meningeal layer, the 'dura mater' ('tough mother'), is quite different. The dura is a dense layer of connective tissue that lies between the skull and the arachnoid mater. It envelops the whole brain like a thick canvas sack. The dura is not solely formed from embryonic neural tissue, but also from the same tissue that goes on to form blood vessels and immune cells. In some respects, it is more like the rest of the body than the brain. The dura's best-known function is to drain fluids away from the brain. In a number of locations, just below the skull, the dura opens up to form hollow channels called dural sinuses. These cisterns collect fluid travelling away from brain tissue, allowing it to empty into veins in the neck on their way back to the heart.

By the early 2010s it was becoming clear that immune cells could in fact travel to, survey, and leave the meninges of the brain.[4] But why? And how? The prevailing theory – of the brain having nothing to do with the immune system – had been strengthened by the fact that the brain didn't house any lymphatic vessels – the immune system's transport system. One of the few who questioned this strange state of affairs – that immune cells could travel to and from the meninges in the absence of lymphatic vessels – was Jonathan Kipnis, then a professor at the University of Virginia. 'We were trying to find out how specific immune cells, T-cells, migrate in and out of the brain,' Kipnis recalls as we speak via video call. 'In the summer of 2014 one of my postdoctoral colleagues, Antoine Louveau, mounted the whole meninges of a mouse to look at it under a microscope.' Louveau observed the thick vessels of the dural venous sinuses forming a Y shape on the base of the skullcap: nothing out of the ordinary. But as he looked closer, it did seem as though there was an abnormally large number of immune cells in the sinus. He magnified, and he magnified again. What he saw was truly bizarre: most of these T-cells were not within the vessel, but ran alongside it. Seemingly suspended in mid-air against the black background of the microscope slide, orderly strips of immune cells followed the course of the dural venous sinus. Could these be veins? No – if they were, then they'd be stained red by the lab marker that identifies blood vessels. 'So then we went to our other colleagues and asked for markers that label lymphatic vessels, just to rule this out,' Kipnis remarks, in a tone that suggested a sprinkling of hope in a broth of scepticism. Everyone knew that the brain didn't have lymphatic vessels, so why look? After the stain was applied, Louveau peered back down the microscope, and this twenty-something postdoc saw something that should not have been there. He beckoned Kipnis over: 'I think we have something.'

Lying alongside the wide dural venous sinuses, something unexpected had appeared. Radiant in the fluorescent lime-green of the new stain was a network of lymphatic vessels. The implication was too astounding to seem plausible: in the twenty-first century we were still discovering new anatomy. Hiding in plain sight alongside much larger structures, these 'meningeal lymphatics' had probably been seen by humans many times before, but never perceived, understood or recorded. We had missed it – all of us – for all of scientific history. 'I actually don't remember my reaction,' says Kipnis, 'but I'm told that there was a lot of screaming, clapping and dancing in the hallways. It was a eureka moment.' Kipnis and his team had been looking for gates that would let immune cells in and out of the brain. What they found was a whole road network. These meningeal lymphatic vessels directly connect the borders of the brain to the rest of the body's immune system. The discovery of meningeal lymphatics was itself huge – but its implications were head-spinning.

A central tenet of scientific discovery is for the results to be replicable. Fortunately for Kipnis, the understandable scientific scepticism that met these findings was considerably dampened when it came to light that a European team had also, independently, discovered meningeal lymphatic vessels in mice, publishing their paper a mere month after Kipnis's.[5] Soon laboratories from all over the world were finding these systems in mice and other animals.[6] But mice are not men, and the next task was to see whether humans have meningeal lymphatics too. In 2017, human subjects were injected with a specific dye that collects in lymphatic vessels, and their brains were visualized in an MRI machine – the images clearly demonstrated a complex network of meningeal lymphatics.[7]

This discovery not only revealed immune highways between brain and body, but also provided a previously missing piece of the puzzle of 'brain drain'. Alongside blood, dural sinuses also drain CSF from the brain – but no one really knew how. This clear fluid that fills the subarachnoid space both cushions the brain from blows to the head and buoys it up, preventing the lower regions of the brain being crushed by its own weight. Until fairly recently it was thought that these were the CSF's only roles. It was also thought that its circulation was pretty straightforward: after being produced in caverns deep within the brain called ventricles, CSF just circulated around the subarachnoid space in the brain and the spine. (You may well have heard of CSF as being the fluid that's extracted from the base of the spine in a lumbar puncture or 'spinal tap'.) It would then pass up into the dural venous sinuses via little pouches in the dura called arachnoid granulations, where it would drain into the blood. But in 2012, Danish neuroscientist Professor Maiken Nedergaard and her team at the University of Rochester discovered that CSF also plays a key role in the brain's waste management.[8] The CSF does not simply buoy the brain tissue; it can enter the brain and give it a wash. Arteries that supply the brain with blood travel through the CSF-filled arachnoid space before diving down into the brain proper. Nedergaard found that as these arteries entered the brain, there is a space between the outside of the artery wall and the brain tissue, making the artery a tube within a tube. This newly discovered outer tube is filled with CSF from the arachnoid space. It was also clear that veins, taking blood away from the brain, are sheathed with a similar outer tube. As the inner tube (the artery) pulsates, it forces CSF from the outer tube into the brain tissue. The CSF then washes through the neurons of the brain, clearing away toxins and waste products before being absorbed into the outer tube of a vein, leaving the brain and travelling into a dural venous sinus. Nedergaard termed this the 'glymphatic system'. This is because it acts in a relatively similar way to conventional lymphatics in the body, which help drain waste products from organs; the 'gl-' of 'glymphatic' comes from 'glia' – the other types of brain cells than neurons, which we will encounter shortly. Specific glial cells – called astrocytes – form the wall of this outer tube and enable CSF to enter brain tissue to wash over it.

Understandably, most neuroscientists don't often dwell on the intricacies of the brain's plumbing system. But the way in which Kipnis's team built on Nedergaard's discovery is critical: they gathered convincing evidence that meningeal lymphatics are the main drains of CSF from the brain.[9] We finally understand how it all fits together: clean CSF is made deep within the ventricles of the brain; it then washes across the brain tissue via Nedergaard's glymphatic system. This dirty CSF then crosses into the dura, a layer rich in immune cells, before draining down the meningeal lymphatic vessels. Crucially, diffused within this dirty CSF are thousands of tiny molecules that provide snapshots of the brain's health to immune cells in the meninges – including whether it is infected or inflamed. This means that while immune cells from the body do not have unregulated access to the brain tissue, they constantly taste the brain's sewage, sampling small proteins and other molecules produced inside the brain.

The picture that science is slowly painting is extraordinary. The border of our brain – the dura mater – is patrolled by a large and diverse army of immune cells, making it similar to other immune-active organs of the body. These cells can receive messages from the CSF leaving the brain tissue – specifically, as we saw in the last chapter, by capturing antigens, substances that the immune cells can identify. They can then travel down the meningeal lymph vessels and present these molecules to specialized immune cells in lymph nodes all over the body. These immune cells can also see what is going on in the rest of the body, by accessing the blood that flows through dural sinuses.[10] A lot of this happens in a dural sinus that runs along your crown, from the top of your forehead to the back of your head – as Kipnis puts it: 'In a war, the top of the hill is always a good place.' Right at the top of your head, just under your skull, sit the battlements of your immune defences, surveying both brain and body. Your immune system has the high ground.

The discoveries have not stopped there. At the turn of the 2020s, Kipnis's team came across another oddity: many of the resident immune cells of the meninges seemed to come from different neighbourhoods to regular immune cells. Most immune cells are born deep within the marrow of large bones, such as the hip or the sternum. Once they have differentiated from a primordial swamp of stem cells, they enter the blood circulation to begin their working life. But it seemed that meningeal immune cells did not share this lineage and had not arrived from peripheral blood vessels, yet they replenished at a similar rate to other immune cells. Where on earth were they coming from? The missing piece to this puzzle had fortuitously been found a couple of years earlier, in 2018, by a team at Harvard. Using incredibly detailed imaging, they found that the skull was connected to the dura by tiny bony structures termed 'skull channels'.[11] These stalagmite-like structures contain vessels, which Kipnis's team found to be the roads that link meningeal immune cells with their source: the skull's own bone marrow.[12] Another discovery by this team, in 2022, found that as dirty CSF washes out of the brain, it can travel up these skull channels, presenting immunological information from the brain tissue to immune cells within the skull's bone marrow.[13] Activity within the brain is constantly shaping the make-up of the skull's bone marrow. Your skull is not just a crash helmet, protecting your brain from the punches, falls and bites of the macroscopic world. It is also an immune watchtower, surveying the brain for microscopic threats.

A decade ago, the area between the brain and the skull was seen as biological bubble wrap, filled with shock-absorbing fluid. Now it is clear that the brain is encased by an intricate immunological organ. The genius of the meningeal immune system is that it can provide constant surveillance of the brain without the need for immune cells from the body to root around in the brain's precious nerves and disrupt their fragile networks. This discovery alone was enough to declare a neuroscience revolution, one that is already showing promise for treating diseases of the brain. But the last decade or so has seen an immunological breakthrough of equal, perhaps even greater proportions. And it involves the brain itself.

In the brain

The brain is made up of billions of neurons, a spaghetti-like tangle of electrical wires that form the networks of our brain and mind.

But when neurons were discovered in the early nineteenth century, it was clear that these nerve cells were not alone. In 1856 the German physician Rudolf Virchow noticed that neurons were enmeshed in some kind of tissue made by mysterious, non-neuronal cells. He suspected that these provided scaffolding to the wires of the brain, calling these new cells 'glia', the word being the Greek for 'glue'. For over a century it was reasonably assumed that glia were passive support cells for the neuronal superstars. Over the past few decades, however, it has become clear that this could not be further from the truth.

Glia are legion. There are roughly as many glial cells as there are neurons: tens of billions. And they are wonderfully diverse. We have already come across the astrocyte, which is a beautiful, star-shaped cell. It uses some of its many long legs to wrap around blood vessels, helping to form the blood-brain barrier. Its other legs wrap around numerous neurons and their synapses, providing them with essential nutrients and modulating the activity across the synapse. Ependymal cells, meanwhile, are responsible for producing CSF and, wafting it with their hair-like processes, gently send the liquid on its way. Oligodendrocytes coat wires of neurons in an insulating sheath. But perhaps the most intriguing of all glial cells are microglia.

Microglia were first discovered in 1919 by the Spanish neuroscientist Pío del Río Hortega, colleague of neuron-discoverer Santiago Ramón y Cajal.[14] These cells, as their name suggests, tend to be smaller than other glia. They look like little octopuses: a central body radiating lots of active tentacles. In the 1920s microglia were observed wandering around the brain – seemingly scavenging around – engulfing and destroying damaged cells within a brain tumour.[15] They appeared to behave like the immune system's macrophages, roving about, surveying their environment with their arms, clearing up debris and destroying damaged or infected brain cells. How they had got into the brain was anyone's guess. By the 2010s the reason behind this became clear. Microglia are not actually glia, in the sense that they are not derived from brain cells in embryonic development. Instead, they derive from primitive macrophages in the embryo and enter the brain before the blood-brain barrier is formed.[16] Microglia are a mercenary offshoot of the immune army, tasked with defending the brain. And this is not a short-term arrangement. Think of the Swiss, who in 1506 promised to provide soldiers to the Vatican to protect the Pope; with microglian faithfulness, the Swiss Guard are still there today.

The more we study microglia, the more we find that they behave like immune cells. Alongside their scavenging and engulfing roles, it turns out that they too produce cytokines and are themselves activated and influenced by these immunological molecules. Once activated by damage or infection, microglia can become increasingly inflammatory, producing pro-inflammatory cytokines and even presenting antigens from the battle site to peripheral immune cells that enter the brain. We used to believe that an influx of the body's immune cells into the brain is a sign of disease, but a remarkable study in 2020 found that in the brain of a foetus, specialized T-cells from the body's immune system travel through the blood-brain barrier and interact with microglia, helping them develop into mature immune cells.[17] These young guards of the brain are privileged to have experienced veterans of the immune system visit and train them up.

The fact that the brain has its own immune army – and one that communicates with the rest of the immune system – may mean rewriting the textbooks. We are only just beginning to find out how this system works in health, let alone its emerging role in disease. But there is another recently discovered and utterly remarkable side to this multifaceted cell. Contrary to what you might think, as you develop from a young child into an adult, the number of neurons in the brain does not increase. In fact, the number of synaptic connections between neurons in your brain peaked when you were a toddler. During late childhood, adolescence and your early twenties, these synapses are whittled down. This is no bad thing: it increases efficiency and helps make you who you are. The brain isn't like a Lego construction, with elements gradually being added on to it; it is instead like a large block of clay that needs to be slowly sculpted by a potter into something meaningful and functional. Neural pathways and synapses that are regularly used tend to stay put; those that fall into disuse are pruned away. As neuroscientist Carla Shatz pithily puts it, 'neurons that fire together, wire together'. Or to put it another way: use it, or lose it. This is the basis of how the brain changes itself and how we learn – a process called neuroplasticity.

In the early 2000s this idea of synaptic pruning was well established, but how the process actually occurred was a bit of a mystery. In 2007 Beth Stevens, a postdoctoral scientist working in the late Ben Barres's lab at Stanford University, found a very unexpected answer. In the immune system, specific immune molecules called complement proteins attach themselves to microbes or damaged cells and act as an 'EAT ME!' signal to macrophages. Beth found that mice that lacked complement proteins were unable to prune their synapses.[18] A few years later, in 2013, after Beth Stevens had moved to Boston Children's Hospital to start her own lab, she and her postdoctoral fellow Dorothy Schafer found that once these unused synapses are coated in complement proteins, microglia turn up to engulf and destroy them (like coating any food in chocolate sauce when there's a young child in the vicinity).[19] Stevens's team later found that the synapses that are in regular use produce a 'DON'T EAT ME!' signal (like covering a child's meal in kale).[20] Microglia are not just the brain's equivalent of macrophages, they are the sculptors of our brains. If microglia are to be renamed – a suggestion made by some, as they are not true glial cells – my suggestion would be 'microangelos'. Although, somehow, I don't see that catching on. Microglia's role in synaptic pruning may also have significant relevance for developmental disorders. Abnormalities in neuronal pruning can be seen in subsets of people with neurodevelopmental conditions such as schizophrenia. While we don't yet know exactly how these alterations in brain connectivity manifest in disease, there is hope that treatments targeting microglia and complement proteins may one day be of great use.[21]

The immune-brain loop

The case against the immune-privileged brain is now overwhelming. Not only is the brain surrounded by a halo of immune cells in the meninges, it is also full of specialized immune cells itself. Study by study, it is becoming abundantly clear that the brain communicates with the immune system all over the body. These signals go in both directions – from body to brain and brain to body – through many different channels. The first hurdle that has to be crossed is the blood-brain barrier, that physical manifestation of Descartes' concept of mind-body dualism. It turns out that this formidable multilayered wall is much more scalable than previously thought. The endothelial cells, which line the inner layer of the blood-brain barrier, display specific receptors that can be activated by cytokines – messenger proteins of the immune system. This acts like ringing a doorbell, with the endothelial cells passing messages to glial cells inside the brain. There are also a few specialized gates in specific locations in the brain called circumventricular organs, where brain tissue can sample small amounts of immune molecules in the peripheral circulation. And, as we know, a cytokine circulating in the blood that reaches the meninges surrounding the brain can find itself being sucked down the plughole of the glymphatic system, entering the brain via CSF.

Another series of bidirectional highways between the immune system and the brain come in the form of nerves. It has long been known that peripheral nerves reach almost every tissue in the body, providing information to the brain and acting on the brain's commands. However, it was also long assumed that this did not include information from the immune system – another assumption that has now decisively been proven very wrong. One brain-immune connection has been termed the 'inflammatory reflex'. Following the triggering of inflammation in the body, inflammatory cytokines are detected by the vagus nerve. *Vagus* is Latin for 'wanderer', and this remarkably long nerve snakes from the brain all the way down to the colon, connecting to almost every internal organ on its way down. The vagus alerts the brain to inflammation in the peripheries, with the signal bypassing the blood-brain barrier. The brain then sends a signal back – via the vagus nerve – to the spleen. This is an unromantic organ, seemingly unloved and neglected down in the back left corner of your abdomen, but alongside filtering the blood it is a major garrison of the immune army. This nerve impulse from the vagus has the remarkable ability to calm the immune cells down, making them produce fewer inflammatory cytokines. This helps reduce the likelihood of an inflammatory overreaction, which could cause unsustainable collateral damage. This response, an immune-brain loop, is an example of homeostasis: the brain and body striving to achieve a balanced, steady internal state – a resetting of the fire alarm.

For the last decade or so it has been evident that the brain and the immune system talk to each other, but one of the most exciting discoveries in this field came in late 2021. A team from the Israel Institute of Technology located the precise neurons in a mouse's brain that are activated when they experienced bowel inflammation after drinking water laced with a chemical irritant.* This was in a part of the cortex called the insula, a fascinating area that mediates the relationship between brain and body, and a region to which we will return. This discovery alone would have been an achievement,

---

* It's important to recognize that most breakthroughs in human health come at the expense of animals. As is often the case in the realm of ethics, there are few easy answers. The studies I include have followed strict international guidelines on animal welfare.

but what the scientists did next was astounding. Once the mouse's inflammation had fully settled, the scientists used a cutting-edge technique called chemogenetics to stimulate those same neurons in the brain. When they did this – only stimulating the brain in an otherwise healthy mouse – they caused the same inflammatory response in the gut: no poisoned water needed. The brain had not just stored and remembered the fact that inflammation had occurred; it remembered exactly where it had occurred, and reproduced the same response with no trigger in the body or in the environment. To complete the circle, the scientists again caused bowel inflammation via the tainted water, but dampened down the immune response by deactivating those same neurons in the brain.

Until recently it was scientifically embarrassing to suggest the existence of a connection between the brain and the immune system. The immune system stopped at the neck. Today it is clear – and this is no overstatement – that the brain is an immunological organ. It has multiple lines of communication with the body's immune army, has a powerful immune garrison in the skull and meninges, and even has its own specialized mercenary force of microglia. The brain and the immune system are tethered together by vessels (lymphatic and blood) and nerves. They speak the same language. You could even say that they are part of the same system. There is a part of us that feels this brain-body link on a deep, instinctive level, but the human mind's need for compartmentalization – which has influenced the development of separate scientific fields and the ordered progression of medical education – has blinded us. The many streams of evidence linking the brain and immune system show that their connection can be no evolutionary accident; it must serve an adaptive purpose. Now that we know these two systems are inextricably linked, we can enjoy finding out *why*.

3

The Sick Sense

How your mind is part of your immune defence

Out, damned spot! Out, I say!

WILLIAM SHAKESPEARE,
LADY MACBETH IN *MACBETH*

Why wasn't I at the ABBA museum, like a normal tourist? The nurse brandished a needle and began to identify its fleshy target on my left shoulder. I had travelled to Stockholm, the Swedish city of picturesque, pastel-painted buildings, wooded islands and a rich artistic heritage. But I had not come for the architecture, the archipelago or even ABBA. I had come to find out *why* we feel ill. More specifically, I had come to find out why our body and brain *wants* us to feel ill when we are infected or inflamed. And, naturally, this involved me being injected with a dose of bacteria.

But, before we explore my slightly masochistic self-experiment, we need to go back a few hours, to the shiny, new offices of Stockholm University's Psychology Department. I was speaking with Mats Lekander, a professor of psychology and immunology at both Stockholm University and the Karolinska Institutet, the institution that awards the Nobel Prize in Physiology or Medicine each year.

Lekander had little interest in the immune system when he started his undergraduate Psychology degree, at Uppsala University; he was set on becoming a child psychologist. One morning, however, an almost throwaway line in a textbook he was reading entered his mind like divine revelation. 'It was about early research suggesting that our immune cells could read messages sent from the nervous system . . . I began to think about whether emotions or behaviours could influence immunity,' Lekander remembers. At the time, very little was known about the relationship between the immune system and the mind, and the textbook didn't expand much further. But the idea gripped him, triggering a cascade of questions, hypotheses and experiment ideas. He described starting to view the immune system as a 'floating brain' that communicated with the brain of our nervous system. And while his initial interest was in how the brain can influence the immune system, he began to look in the other direction, at how bodily processes in the periphery influence your thoughts, feelings and behaviours. 'I became fascinated by how the immune system needs help from behaviour. Because the threat of microbes is and has been so great, behaviour needs to be recruited to keep you healthy.'

That stopped me in my tracks. The immune system recruits behaviour? This went against all my preconceptions of the brain being an all-controlling, sovereign master that bosses the body about. 'Of course the brain is a regulator,' Lekander explained, 'but it is also shaped and regulated.' This makes sense when you realize that the body has not evolved primarily to support a thinking brain; the whole human organism has instead adapted to successfully survive and thrive in a hostile environment. Lekander argues that 'we have one integrated defence system that is made of different, specialized components that are good at detecting different threats at different levels.' For example, the classical senses of the nervous system – such as sight, sound and touch – protect us from 'macro' threats to our body, from a lion's jaw to the blade of an enemy's sword. But, throughout human history, our greatest enemy has come in 'micro' form. Most deaths in human history have been at the hands of organisms we can't even see: bacteria, parasites and viruses. The mind-boggling complexity of our immune system chronicles a long history of an arms race against invisible pathogens. This defensive system, however, does not work on its own. Its greatest ally is our brain, which it calls in to give a helping hand. So, what behaviours does our immune system need us to perform – and why?

All of us have experienced 'sickness behaviour'. This is the constellation of symptoms and behaviours we exhibit following infection or inflammation: tiredness, low mood, loss of appetite, increased pain sensitivity, slower movements, a lowered drive to explore the outside world and a reduced ability to experience pleasure. It's an overwhelming, motivating desire to retreat and turn in on ourselves. Think about the last time you were subdued by the flu, or Covid. You were probably curled up under a blanket watching Netflix, rather than preparing your CV, embarking on a cryptic crossword or organizing a dinner party. It is completely extraordinary that scientists and clinicians have, until recently, given little thought to the links between the immune system and the mind when it comes to getting ill, considering that most of the real-world symptoms of immune activation involve mood and motivation.

If the brain and the body are intimately connected, we are only just beginning to decode their language and listen in on their conversations. Over the past twenty years or so, pioneering scientists like Lekander have tried to explore these links by eliciting sickness behaviour in healthy subjects. One way of doing this is giving them a vaccination, which can bring about a short-lived inflammatory response. Although, *can* is an operative word – not all vaccinations (indeed, not all infections) reliably result in sickness behaviour. Lekander tells me how frustrating it is that many of his volunteers haven't produced a response worth studying. And once, it was embarrassing: a radio presenter visited his lab and was given a vaccination two hours before a live interview with Lekander in which he proudly declared, 'No, I don't feel a thing ... I feel absolutely fine!'

So, in his search to find a reliable way of inducing sickness behaviour in healthy subjects, Lekander decided to go for the strong stuff: endotoxin. Endotoxin (also known as lipopolysaccharide, or LPS) is a slightly ominous term for a large molecule, made up of a lipid and a long tail of sugars, found in the outer membrane of some bacteria. This is detected by pattern-recognition receptors on immune cells (which we came across in Chapter 1), which in this case are known as toll-like receptor 4, or TLR4. When TLR4 detects endotoxin, a powerful inflammatory response is activated. Endotoxin is so potent that sickness can be induced by a couple of nanograms – a few billionths of a gram; so much so that endotoxin is often a cause of septic shock, which can be fatal. Understandably, Lekander had to provide a lot of evidence to the Swedish authorities to persuade them that giving endotoxin to healthy volunteers was safe and worthwhile. His team of psychologists made an unlikely alliance with a group of hospital doctors specializing in sepsis and hospital infections, and together they worked out a dosage and a model that has now been used successfully many times in labs across the world. This included testing it on themselves. 'Actually, I'm planning a social endotoxin party with my friends soon,' Lekander adds. I didn't ask whether he was serious, but I'm fairly sure that he was. Sometimes progress requires a few milligrams of madness.

Endotoxin has proved a great way of studying how inflammation affects the mind. When you take it, your mood drops and anxiety levels increase. When subjects are given endotoxin, they tend to feel more socially disconnected,[1] are less likely to seek rewards[2] and are worse at reading emotions in others.[3] Imaging of the brain strongly supports the behaviours seen in these experimental settings: on scans there is a noticeable overlap between brain changes in endotoxin-induced sickness and in depression.[4] Powerful new imaging techniques are also revealing how endotoxin-induced peripheral inflammation can rapidly lead to inflammation in the brain itself: neuroinflammation.[5] But while sickness behaviour is miserable, Lekander makes sure to point out his view that it is both adaptive and advantageous to humans. If our outer defences are breached by a pathogenic bacterium or virus, by resting and isolating ourselves from others we 'save energy for the hugely energy-consuming process of fever and immune activation, reduce our risk of being hunted down by predators and reduce the chance of spreading an infection around our tribe'. This is a crucial point. It can be easy to assume that the behavioural symptoms associated with infection are a sign of weakness, damage and poor coping. But, in the short term, it is actually a healthy, beneficial defence mechanism for us and for others. If the brain was completely shut off from the immune system, humanity may not have survived its pathogenic wars.

After our discussion, I felt that the best way to reflect on sickness behaviour would be to experience it myself. Lekander – with the curious and slightly pitying look of someone having to help an Englishman make himself ill – encouraged me to start with the typhoid fever vaccine. This tends to be weaker than endotoxin, and doesn't require a full day of medical monitoring. The typhoid fever vaccine contains a small part of the bacterium responsible for typhoid: *Salmonella enterica*.* It consists of bits of the bacteria's distinct outer capsule – something that would sufficiently identify itself to my immune system without the risk of me contracting the full-blown disease. Over the course of a few days, my adaptive immune system would form a memory of the bacteria, ready to mount a swift response if I were to be infected with a live pathogen in the future.

So it was that I found myself in a Swedish travel clinic, belatedly

---

* Specifically *Salmonella enterica* subsp. *enterica* serovar Typhi. Essentially, the disease-causing bacteria are a specific subtype (also known as a serovar) of the *Salmonella enterica* species.

regretting the life choices that had led me there. Too late: after the painless injection, given by an efficient yet reassuring Swedish nurse, I left the clinic to head to dinner. I was to meet my wife and two friends who had travelled with me to Stockholm. As I strolled down the cobbled streets of the Old Town, nothing seemed out of the ordinary. Streams of cyclists flowed out towards the suburbs. But, already, the wheels of my immune system had started to turn. Cells of my innate immune system – the front line of our fight against microbes – had already detected molecular patterns on the pieces of *Salmonella enterica*, using their pattern recognition receptors (PRRs). As we saw in Chapter 1, PRRs are essentially bacterial barcode-readers, encoded in our genes and provided to most of our immune cells so that they are always equipped to detect common microbial patterns. The detection of a bacterial pattern triggers reactions in the immune cell that will eventually lead to an inflammatory response. This includes the production of cytokines such as interleukin-6 (IL-6).[6] These hormone-like molecules create and regulate immune responses around the body and – crucially – around the brain. As I reached the restaurant to join my wife and friends, inflammatory cytokines were already travelling through my bloodstream and beginning to communicate with my brain.

The meal was delicious: a fish soup with its catch fresh from the Baltic. My friends regaled me with stories of their sightseeing – from royal palaces to coffee breaks in trendy cafes – laced with the self-deprecating accounts of the many faux pas made by Brits abroad. But once they asked me about my interview at Stockholm University and my subsequent injection, I realized that I couldn't drum up either the words or the enthusiasm to tell them anything. This is unusual: given a captive audience, I can talk science for hours – whether said audience has lost interest or not. I now had an odd but overwhelming feeling that I no longer wanted to be around my friends. Indeed, I was developing an aversion to being near anyone. I excused myself and retreated back to the hotel. When I reached the bedroom, the sickness behaviour really began to kick in. I sat down at the small desk by the bed in an attempt to record how I was feeling, but would lose my train of thought as soon as I started typing. Reading a book was also impossible with a non-existent attention span. I soon had full-blown anhedonia: the inability to feel pleasure. Even the simplest dopamine-bomb TV show had lost its lustre. I felt morose, allowing my mind to entertain uncharacter-istically defeatist thoughts: what was the point of this painful self-experiment? Would anyone even be interested in the results? By this point I had curled up into a ball under the sheets. I had physically and mentally withdrawn into myself. It didn't take long, however, for fatigue to overwhelm me: I fell asleep at around 8 p.m.

The effects of the vaccine had mostly subsided by daybreak, although I had no desire for the ritual morning hug with my wife and no appetite for breakfast. The very thought of milk made me nause-ous. I was also a bit grouchy. To my relief, and to those around me, the symptoms resolved by around lunchtime. I could finally enjoy the sights of Stockholm and reflect on my chat with Mats Lekander the day before. Something he had said was playing on my mind: 'Sickness behaviour is part of the immune defence, and it is mainly activated when pathogens are detected in the body.' I had to clarify why he said *mainly*; surely immune-related behaviours can only be triggered by the detection of a pathogen in the body? Surely the immune system needs to be activated by the presence of 'physical' pathogenic mater-ial? But as I arose from my inflammatory torpor I remembered that Lekander had made it clear that the immune system is much more intelligent than I was giving it credit for: incredibly, it recruits behav-iours to avoid being infected in the first place. Think of the way we are repelled by objects that have a high risk of carrying pathogenic microbes: from faeces to rotten food, and even other humans exhibit-ing sickness behaviour. From an adaptation perspective, this also makes sense. Sickness behaviour comes at a great energy cost, so not getting infected at all has a considerable survival advantage.

Avoiding infection requires first identifying those who could be harbouring an infectious agent. Remarkably, although it makes intuitive sense that healthy individuals can detect those who are sick – and thus try to avoid them – this was not explored scientifically until very recently. In 2018 Mats Lekander's team found that untrained people are adept at detecting whether a subject has been injected with endotoxin compared to a placebo, just by looking at a photo of the subject's face two hours after injection.[7] When asked to describe the differences, they put it down to subtle but noticeable changes, such as paler lips, a droopier mouth and less glossy skin. They also found that people are able to detect experimentally inflamed individuals from healthy controls, just by observing videos of them walking.[8] The changes may be subtle, but humans are exquisitely attuned to detect and analyse biological motion. The team later conducted a thorough gait analysis, which identified a pattern of changes in the people who were inflamed: 'shorter, slower and wider strides, less arm extension, less knee flexion and a more downward tilting head while walking'.[9] Perhaps it is no surprise that this is how we tend to depict zombies: a monster that stems from a collective, primal human fear of the infected.

We can see sickness, but vision is not the only sense engaged in avoiding infection. When it comes to our sense of smell, we humans sell ourselves short. Granted, we don't seem to rely on our relatively small snouts to detect our world as much as our canine companions do, but we underestimate olfaction at our peril. When I caught Covid for the first time, in April 2020 (an inevitability considering that I and my fellow front-line NHS staff were given utterly inadequate PPE – but I digress), I completely lost my sense of smell, and with it my sense of taste, for about nine months. I soon became acutely aware that I couldn't tell whether food or milk had gone off, and had to rely on others to act as a detection system. It would be intriguing to see, though, if we can somehow smell infection or inflammation. Dogs certainly can. In 2012, researchers from Amsterdam trained a two-year-old beagle to lie down on the floor if it could smell *Clostridioides difficile,* the bacterium known for causing colon inflammation and diarrhoea.[10] The dog detective was taken to three hundred patients across two hospitals and stayed standing in the presence of 98 per cent of non-infected patients but lay down when it was alongside 83 per cent of the infected individuals. Lekander wanted to find out whether humans could do something similar. As he outlines in his book *The Inflamed Feeling,*[11] he was partly inspired by the ancient physicians, who, without the tools of modern medicine, smelled and sometimes tasted the urine of their patients to arrive at a diagnosis. Thomas Willis, the brain-dissecting seventeenth-century English doctor we met in Chapter 1, coined the term 'diabetes mellitus', as patients' urine would smell 'wonderfully sweet, as though honey were mixt in it',[12] 'mel' being the Latin for honey, and 'diabetes' referring to the discharging (literally 'siphoning') of excessive amounts of urine. In 2014, Lekander teamed up with olfaction experts to carry out one of my favourite types of study (albeit largely for their comedic value): a sweaty T-shirt study.[13] Participants were injected with either endotoxin or a placebo and then had to wear a tight-fitting T-shirt for four hours. Deodorant was not allowed; their body odour needed to stay *au naturel.* The shirts were then given to another (rather unfortunate) group who had to sniff the armpit areas and rate the intensity and quality of their smell. On average, they rated the odour of the endotoxin-inflamed participants as less healthy and less pleasant than the controls'. The Swedish team also found that even if individuals were not consciously able to detect differences between body odour from sick, endotoxin-exposed people and controls, the smell of the sweat of the sick significantly increased activation in networks in the brain responsible for odour perception.[14] The researchers also found that individuals are able to discriminate healthy controls from recently inflamed subjects just by the smell of urine or body odour, and they are more likely to show dislike towards inflamed subjects.[15]

All in all, we are well attuned to use multiple streams of sensory data to detect whether others are sick, with or without our subjective awareness. Crucially, we are motivated to avoid the sick, too. Studies have clearly shown that viewing images of sick humans evokes intense feelings of disgust and changes our body's physiology by both raising body temperature and priming an immune response.[16] Simply seeing someone exhibiting sickness behaviour can activate your defence system, changing immunity and behaviour. Reflecting back on my typhoid vaccine experiment, it's clear that illness avoidance was also at play. What I failed to mention earlier was that my lack of desire to give my wife a morning hug was strongly reciprocated: 'you look gross' are the words she used. When our immune system kicks into gear in the face of an infectious threat, it's almost as if we become magnetized: drawn towards focusing on our inner selves and repelling others. But others also pick up on what is going on, and are repelled from us. Avoidance goes both ways.

This repulsion is not usually isolated to two individuals; it ripples through society. The briefest examination of human history will bring up various examples of how the identification of the infected – with the subsequent disgust and avoidance it produces – has led to stigma and suffering. For thousands of years (and, sadly, until very recently), leprosy sufferers were commonly housed in 'leper colonies' far away from population centres. Untreated leprosy is both infectious and deforming; perhaps it was this combination that led to it being particularly feared. The fact that syphilis was called 'The Italian Disease' by the French, 'The French Disease' by the Italians, 'The Polish Disease' by the Russians and 'The Christian Disease' by the Turks shows how infection can further isolate and ostracize an outgroup. More recently, the moral panic following the emergence of HIV showed how quickly walls of fear could be built, separating sexualities and races. And, of course, the Covid-19 pandemic has revealed the xenophobia that so often accompanies the fear of contagion.

It's clear that social distancing is baked into the human condition, and while this slows down the spread of infection, it comes with ugly social side effects. On the one hand, it's remarkable that our immune system and brain have developed behaviours to avoid infection, thousands upon thousands of years before humans had any explicit knowledge of the microbial world. But that also means that disease avoidance is rather crude and non-specific. Infection avoidance is a powerful instrument, but a blunt one. Of key importance is a phenomenon I will term 'infection bias'. In trying to detect pathogens, our body's defence system has to make a trade-off between two possible types of error. A 'false positive' occurs when we incorrectly classify a non-infectious cue as infectious – we might avoid a person or group of people we believe to be diseased, even if they pose no threat at all. A 'false negative' is incorrectly classifying an infectious cue as something safe – we eat the infected food or kiss the infectious human. Across millennia – in a world without antibiotics, antivirals or vaccines – the human defence system learned that false negatives are much more costly than false positives. Better safe than sorry. We have, therefore, an immune-brain defence system that leans towards being overprotective. It can overestimate the severity of disease in our own body and overgeneralize disease threats in the outside world, particularly during pandemics. The psychological scientist Mark Schaller – who has pioneered this field of infection-avoidance behaviours, terming them 'the behavioural immune system' – found that healthy participants were more likely to conform to social norms if they felt at risk of catching an infection.[17] When people are aware of the threat of infectious disease, they are more likely to judge negatively those who break social and moral norms,[18] as well as becoming more hostile towards outgroups, such as immigrants and minorities.[19] Just as an individual retreats inwards and avoids others when the immune system detects a threat, so a society detecting real or imagined infection becomes more insular.

This link between the social and the immunological is strengthened by evidence demonstrating surprisingly little distance between
physical and moral disgust, which are both processed in the same
area of the brain. In one study, participants watched a movie clip of
something disgusting (for those who really want to see it, look up
'the worst toilet in Scotland' from the 1996 film *Trainspotting*).
Afterwards, half of the participants were asked to wash their hands,
before all participants answered a questionnaire about ethical sce-
narios.[20] Those who had washed their hands after the movie scene
found specific moral actions to be less wrong when asked to evalu-
ate them. So, Lady Macbeth and Pontius Pilate might genuinely
have been cleansing their consciences while washing their hands.
We find it surprisingly difficult to tell the difference between physical and moral hygiene, just as we are often oblivious to how it
influences our behaviour. As we have seen, the line between the
physical and the social becomes increasingly blurred during disease
epidemics and pandemics, but this also occurs in any case of
increased vulnerability to infection. In the first trimester of her first
pregnancy, my wife found that certain characteristics of food – such
as the lumpy textures of most soups and the tangy scent of my
homemade sauerkraut – made her feel intensely nauseous. Heightened disgust sensitivity in pregnancy is likely to be a behavioural
compensation for a temporarily suppressed immune system; the
flexible neuro-immune defence system is trying to further decrease
exposure to infection.[21] Curiously, there is some evidence that this
heightened disgust in pregnancy may be associated with increased
xenophobia and favouritism towards the in-group.[22] The lines we
draw between the biological, the psychological and the social are
much more porous than we assume.

These are quite direct and overt ways in which the immune system
influences social interactions. But, at the end of my discussion with
Mats Lekander, he left me with something that made me realize
that our immune systems may be shaping society in unseen ways,

most of which we have not yet discovered. He told me about the PhD thesis of his student Marta Zakrzewska, who had successfully defended it just the week before. Zakrzewska's work, 'Olfaction and Prejudice', explores how sickness-avoidance behaviours impact society in unexpected ways. She found evidence to show that 'how easily one gets disgusted by body odours is reliably related to negative attitudes towards others'.[23] Zakrzewska found that people who were most predisposed to body-odour disgust were more likely to exhibit explicit prejudice towards a fictitious refugee group, as well as harbour more implicit biases against real-life out-groups.[24] These patterns were reliably reproduced in a number of different countries and cultures across the globe. In her thesis, Zakrzewska concedes that prejudice is, of course, not simply a function of our sense of smell, but these findings are an example of how many of our social biases 'can at least partially be traced back to these primitive disease avoidance functions'. Disgust and avoidance are, respectively, emotions and behaviours adapted to aid our avoidance of pathogens, but they have surprising social consequences. We assume that the social reality we have constructed in the world – philosophy, politics, power – is built by human intellect, reason and will. But we are beginning to find out that it is profoundly influenced by feelings bubbling up from our body's battle with the microbial world. It is intriguing to think that many of the behaviours that have shaped human history may have actually been the collateral damage of an ancient, ongoing microbial war.

Your immune system is not just brain-altering: it is mind-altering. And it needs to be. Your brain and immune system combine to form a unified defence system that helps you avoid getting infected, and helps you recover if you do. On a societal level, human immune systems drive powerful behaviours that break the chain of infection. The most successful pathogens are those that can spread between humans before our immune systems can talk to, and recruit, our minds (SARS-CoV-2, I'm looking at you). Jonathan Kipnis, the scientist who has done so much to reveal the anatomical and cellular pathways linking the immune system and the brain, goes as far as to argue that 'the defining role of the immune system is to sense microorganisms and deliver the necessary information about them to the brain'.[25] He calls the immune system 'the seventh sense'. The five classical senses are sight, smell, taste, touch and hearing. The sixth is contested by senses coming from within the body, such as proprioception (balance) and information coming from our internal organs. To avoid mathematical mix-ups, I just think of the immune system as the 'sick sense'.

It is clear that over just a few short years, centuries of dogma have been turned on their head. We now know the brain and the immune system are tightly enmeshed, and this enables the immune system and the mind to achieve the common goal of surviving and thriving in a microbial world. Much more than close friends who occasionally rely on each other, we are about to see how our mind and our immune army are part of one integrated system. This is not just interesting: it will change the way we view health, treat disease and understand ourselves.

4

A Tale of a Supersystem

A new philosophy of mind, body and world

It is futile to do with more things that which can be done with fewer.

WILLIAM OF OCKHAM,
SUMMA LOGICAE, C. 1323

I am, therefore I think. Does that phrase sound strange to you? You are probably much more familiar with René Descartes' famous dictum 'I think, therefore I am'. This chapter will follow two threads that come from your reading of that first sentence. The first is your experience of reading the phrase. You probably felt a mild sense of intrigue, as well as effort on your part to process the reversal of a familiar phrase. This hints at a curious fact being explored by neuroscientists and philosophers alike: your brain is a prediction machine, and predictions profoundly affect the way you experience the outside world, including what goes on inside your body. The second thread is the content of the sentence. As we discussed in Chapter 1, Descartes' conception of the mind and body as utterly separate entities has seeped deep into the subconscious of Western medicine and society. Many people live as though all of one's essence – thoughts, feelings and personality – is located in a disembodied mind, with the body relegated to the role of a fleshy robot that has evolved to look after the thinking brain. But, as we will see, this could not be further from the truth. The evidence we will encounter in this chapter has profound implications not only for the way we understand how the human organism works, but also how we view illness. We now know that the immune system and the mind are intimately linked on anatomical, physiological and clinical levels. It's now time to see how they are part of the same system on a conceptual level. Join me in hammering the last nail in the coffin of mind-body dualism. Let's start by questioning perception.

The predictive brain

Perceive your world. Look around you and take in your surroundings. Now close your eyes and listen in to the soundscape of your setting. Next – if it's socially acceptable – smell, taste and touch elements of your immediate environment. It seems, compellingly so, that your senses are faithfully and effortlessly recording reality: your eyes a high-tech video camera, your ears an exquisitely sensitive sound-recording device.

But what if this isn't the case, and everything that you experience is essentially a controlled hallucination? This hypothesis – that your brain is not passively receiving and recording information but instead actively constructs and selects your perceptions – is most commonly termed 'predictive processing'. It is a radical, counter-intuitive theory that aims to unify perception, action, thoughts and emotions. Quite an aim and quite a claim. But the scientific, medical and philosophical communities are finding it increasingly persuasive.

At its core, predictive processing posits that our brain generates a model of the outside world, which it constantly builds on and updates. This model of the world aims to predict and explain the torrents of incoming information bombarding our sense organs, only updating if there is a difference between our brain's 'priors' (expectations, theories and beliefs about the world) and sensory input. This difference is called a prediction error. Prediction error is the mismatch between your brain's predictions and reality, between the model and the data. It is, in essence, surprise.

As the brain is a very effective predictive machine, it is hard to shatter the illusion, but what if I told you that the intentional error in this sentence wasn't an oversight by my editor? What you just perceived was not reality, but instead your brain's inference – its best guess – of what it expected to see. This isn't wild guessing; it's a prediction based on mountains of prior evidence and experience. Another effective way of looking behind the curtain of our perceptive machinery is through illusions. Below (Figure 1) is one of my favourite visual illusions: Adelson's checker-shadow illusion. Take a look at squares A and B. You will, without a doubt, perceive square A to be darker than B. But it isn't. If you follow the connecting lines in the second image it becomes abundantly clear that both squares are, in fact, the same shade of grey. You unfailingly see the squares as different shades because the shadow coming off the cylinder convinces your brain that there is a source of light to its right, and that all the squares in the shadow should be darker than surrounding squares. Think of the 2015 viral sensation of 'The Dress': was it white and gold, or blue and black? It depends on your brain's prediction of the lighting conditions. Illusions like these reveal a profound truth about perception. The purpose of the visual system is not to paint a direct picture of reality, but to find the causes of sensations. Your brain is less focused on reality than it is on *meaning*.

This might seem rather complex and illogical; it would be reasonable to ask why the brain should work in this way. Predictive processing is actually a rational, clever and extremely energy-efficient way of running a processing machine. The brain is entombed within a silent box, with no direct access to reality; it can only detect the effects of the world through the senses. The brain then needs to

Figure 1: Adelson's checker-shadow illusion

make sense of the overwhelming data 'noise' emanating from the world. It is much more efficient if it only needs to respond to changes in information compared with what it already knows. Predicting and adjusting is better than constantly reacting.

The concept of predictive processing derives from predictive coding, the process of compressing audio files so that they can be stored and sent without losing resolution and going fuzzy. It's also how a JPEG works: a pixel usually predicts the nature of its closest neighbours, and the differences in the image are found only along clear boundaries between objects. The code can be compressed by coding only the *unexpected* data: the only thing transmitted is the prediction error. To use vision as an example: when light hits our retina, our brain has already created a prediction of what it thinks it will see; it is only errors in our brain's predictions that are passed on to higher levels of the brain.[1] If you are used to an uncluttered kitchen, a sock left on the floor after the laundry's been done would quickly violate your brain's predictions of the kitchen and jump into your field of perception. This concept is consistent with the long-known anatomical fact that there are far more fibres travelling down from the visual cortex (the area of the brain that processes visual information) than travel up to it; this smaller signal only carries the *difference* between expectation and sensory information.[2] Otherwise, we see what we expect.

Let's find a metaphor to help explain this sorcery. Think of your predictive brain as a scientist. This scientist has a theory (our brain's generative model of the world) and they are constantly sampling data from the environment (sensory signals) to refine that theory. To put it simply: perception is hypothesis-testing. When we are faced with new evidence – this prediction error – we update our beliefs. This is neat and energy-efficient: when the scientist arrives in their lab each morning, they don't have to hurriedly swot over all their old undergraduate textbooks or repeat all of the experiments they have done up to the present. They have already formed a broad understanding – a working hypothesis – and the experiments they carry out today will either reinforce this model of the world or challenge it. In a similar way, instead of the brain frantically having to re-interpret the medley of sensory information it receives every millisecond, it only needs to spend energy on significant differences between what it expects to perceive and what is actually out there in the world. In summary, one of the brain's primary roles is to minimize prediction error.

It's one thing getting your head around how predictive processing explains perception; this theory's account for our actions and behaviours is downright spooky. With the tip of one of your index fingers, touch one of the words in this sentence. Before you moved – the split-second you imagined your finger touching the page – your brain decided that your finger was in the wrong place, and initiated movements in your hand and arm to minimize the prediction error. The theory of predictive processing posits that whenever we are faced with prediction error, we can try to minimize it by either updating our model of the world – changing our beliefs – or changing the world to fit our model – through actions and behaviours.

In essence, your brain is a prediction machine that is constantly trying to minimize surprise. But the brain is not just doing this for its own entertainment; it is all in service to the body. Almost every element of our being is constantly trying to stay within a balanced, 'Goldilocks zone' of health, whether it be body temperature, blood-glucose or fluid levels. This thermostat-like seeking of a steady internal state – common to all living organisms – is called homeostasis. What is particularly interesting about organisms with perceiving brains, however, is that they can predict the energy needs of the body in advance and act accordingly. While a simple amoeba or bacterium floating on the surface of a pond eats up whatever is available in its environment, humans book a table for dinner. The brain is not primarily there to think; it is to serve a complex and fragile body in an ever-changing world. 'I am, therefore I think', may be a more apt way of viewing how the brain works. This playful inversion of Descartes is a dictum used by many of those in the predictive processing field, including the British neuroscientist Karl Friston, perhaps the theory's foremost proponent.

When I first came across predictive processing theory, my immediate rebuttal was the fact that humans always seem to be seeking novelty and uncertainty. The very etymology of reading a 'novel' or the 'news' suggests that we like to be surprised. But on further reflection, predictive processing theory might offer a reasonable explanation for short-term novelty seeking: it is actually a weapon in our ability to adapt, survive and thrive in a chaotic and dangerous world. In pushing ourselves outside of our comfort zones, in training for sports and play-fighting, in seeking out the 'safe threats' of horror movies and sampling the danger of extreme sports, we are training our brain and body to minimize future prediction error. Perhaps part of the success of humankind is its ability to predict and prepare years in advance. In sum: you are a chaos management system.

All of this discussion of perception so far has concerned predicting and perceiving the *external* world: 'exteroception'. But we are also able to perceive what is going inside our own bodies, known as 'interoception'. Using the analogy of the perceiving brain as a research scientist busy sampling data from the world and updating their model, the scientist in the department of exteroception is lucky. She has top-of-the-range lab tools such as eyes and ears. On the other hand, the scientist in the interoceptive department has wound up with a paltry research grant. The tools he has for perceiving the internal workings of the body are blunt, though perhaps mercifully so – you don't want to feel food squeezing through every corner of your intestines; perceiving every heartbeat is simply unnecessary, or even a distracting hindrance from a survival standpoint. We all have experienced examples of interoception's inaccuracies. When I was working long stretches of night shifts as a junior doctor, I always put on a little bit of weight. Perhaps this is because the feeling of tiredness is often paired with hunger, both reflecting low energy states. In trying to predict what my body needed, I'm sure that my brain regularly mistook tiredness for hunger. But it's important to remember that interoception's lack of sensitivity doesn't detract from its importance: the sense of hunger is clearly crucial for survival.

Feelings that stem from inside your body – such as from your heart or your intestines – are your brain's best guess of the state of your internal world. Just like exteroception, interoception is the predictive brain trying to make meaning from the mess of sensory information. Your body is a bag of evidence. A tantalizing hypothesis – one that is now firmly buttressed by scientific studies – posits that the brain's interpretation of these sensations results in the experience of emotion. Each emotion we feel is a unique, active construction within our brain's generative model of our inner and outer world, not a discrete entity that is 'triggered' by stimuli.[3] The brain's role in ordering these feelings means that we experience as many, or as few, emotions as we have words or concepts to describe them. Psychology professor and science communicator Lisa Feldman Barrett is one of the main proponents of this theory. She describes emotion as 'your brain's creation of what your bodily sensations mean, in relation to what is going on around you in the world'.[4] She argues that emotions are a 'prescription for action', specifically for behaviours that will keep our body in balance. Emotions are goal-directed, the ultimate goal being the minimization of uncertainty and a balanced state of homeostasis in the body.

Researchers and clinicians in the fields of psychology and psychiatry are increasingly viewing mental health conditions through the lens of predictive processing. Looking at the extremes, psychiatric diseases can be seen as disorders of inference. Over-predicting and attributing excessive salience to sensory data results in hallucinations and delusions – the fantasies that our brains are constantly generating are not kept in check by the sensory evidence that should temper them. The opposite can result in dissociation – the experience of detachment from your emotions, your body or the world. According to the predictive processing framework, what we term mental illness is the formation of a maladaptive model of the world, one ceasing to be updated by prediction errors and inflexible to changing contexts. The predictive brain also explains why humans are so susceptible to 'confirmation bias': it's easier to receive information that confirms our model of the world. Let's say, for example, that you have depression fuelled by a core belief that you are an abject failure in all areas of life. Over time, your brain turns up the gain for evidence confirming this negative belief, and does not pay adequate attention to contrary evidence suggesting that you actually do have strengths and are valued by others. These beliefs also result in behaviours, such as social avoidance, that further strengthen a belief that you have no friends and nothing to offer the world. This becomes a self-fulfilling prophecy: down and down a vicious spiral you fall.

Context is also crucial. A mental health condition may result from what was once a successful, adaptive response that has been transplanted into an unsuitable environment. I once assessed an army veteran with post-traumatic stress disorder (PTSD) who, a decade after returning from a busy tour in Iraq, instinctively jumped off a road bridge in response to the sound of a backfiring car. Thankfully, the waters of a canal below broke his fall. That was a well-adapted response for war-torn Basra, but one unsuited to an uneventful, middle-class suburb of Birmingham. Following a traumatic event – whether it be war, childhood abuse, severe pain, a nasty infection – our brain's predictive baseline can be dislocated and end up becoming hypersensitive, expecting to be in constant danger even long after the peril has passed. Essentially, if someone's model of the world does not successfully adapt to their environment and minimize uncertainty, the result is misery.

Immunoception

Right now you might justifiably be wondering where the immune system comes into all of this. In the last chapter we met what I call the 'sick sense', which describes the feelings that stem from the brain's detection of infection, as well as the behaviours these feelings drive. This form of interoception has been called immunoception. The brain can powerfully perceive infection: think about the depression-like state that so often accompanies the flu. Like all interoceptive feelings, the sensation of the body being infected is not nearly as accurate as exteroceptive senses such as vision – think of our lab scientists with their different research budgets. To illustrate this further, let me share a couple of personal examples of how immunoception can be liable to error and misattribution.

The first date was a roaring success. I'd only briefly chatted to the girl at a house party a few weeks earlier, but I had managed to glean a couple of key facts about her. One was that she studied French at university and had lived in Paris for a year. So I booked Oxford's fanciest French restaurant. The meal was delicious and we instantly clicked. She was impressed enough; I secured a second date, then a third, and before I knew it we were two years into dating. That's when she finally told me her real perspective on the first date. Throughout the meal, she had felt unusually nervous, sensing a high heart rate and a vague sensation of butterflies in her stomach. As the date went on this sensation had become increasingly positive . . . could this be love? It was only when she arrived back at her flat later that evening that the butterflies morphed into caterpillars of nausea, and a deep malaise overcame her. She felt intensely ill, and as she had recently travelled to a malaria-prone region of Africa, made her way to the emergency department of the nearest hospital, where she was treated for gastroenteritis. My date had mistakenly attributed danger messages from her immune system to feelings of attraction. It worked for me, though, as said date is now my wife.

Fast-forward a few years to me waking up with a sore head. I was due to be in work in an hour. I was on a ward at the local hospital, and another wretched wave of Covid was washing its way through patients and staff. Now that I thought about it, my throat did feel ever so slightly sore, and I was exhausted. The more evidence I attributed to having Covid, the worse I felt. I slunk downstairs and took a lateral flow test. Negative. Another: negative. As soon as I was faced with this evidence, and considering the likely alternative – that I don't recover from two large glasses of wine with dinner as well as I used to – I started to feel considerably better.

These are two stories of extremes: in the first case my wife mistook infection for infatuation, in the second case I felt an infection in its absence. Our perception of being ill is not a simple reaction to being infected by a pathogen, like an alarm being set off by the movement of a burglar. Our perceptions are a product of both the bottom-up (sensations brought about by a bacteria or virus) and the top-down (our brain's predictions of the body and the world).

Perceptions are also hugely influenced by context, which itself is a mélange of bottom-up and top-down influences. All human experience is found at the meeting point between our internal model of the world and external reality. This is true for any disease that involves perception. Could you have a virus causing damage in your body and battling your immune system without your awareness? Yes. Could you be experiencing long-term symptoms of illness based on your brain's incorrect predictions that you are still infected, long after the virus has been eliminated from your body? Also yes. Your brain is constantly interpreting your body, just as it does your surroundings. Sometimes this appraisal is very accurate; sometimes it is completely misguided. Most of the time it is somewhere in between.

Perhaps the key message when it comes to appraising our perceptions is to keep an open mind. This sounds painfully obvious, but it often goes against the beliefs ingrained in us by a dualistic society, one that has built a Berlin Wall between body and mind. As we've seen, in being a predictive machine, the brain is also a purpose machine: it is constantly trying to find meaningful, causal explanations for sensory stimuli that fit in with its idea of the world. In an attempt to find causes for diseases or distress, we need to steer away from two fallacies. The first is 'mind over matter' – believing that the symptoms of infections are 'all in your head', or that the damage caused by pathogens can instantly be cured with a positive outlook. The second is a reductionist 'matter-over-mind' approach: all feelings and symptoms must be an accurate reflection of what is going on in the body. In later chapters we will encounter some exceptions: for people driven into deep psychosis by antibodies attacking their brain, a change of mindset isn't going to make much difference. We will also see how psychological stress can derail the immune system in the absence of an infection. Most cases, however, are a dialogue between the brain's model of the world, the senses, and the reality of the world. If we strive to see beyond the binary categories of 'physical' and 'mental', we are equipping ourselves with a toolkit for healing. If our experiences are formed at the confluence of top-down and bottom-up, then effective treatments will come from both directions, too. Mind and body are not separated substances but two halves of a constantly spinning wheel, so healing unbalanced brain-immune interactions will often need to involve effort from both ends.

Unity

We have seen that the brain and the immune system are tethered together anatomically and physiologically. We have seen that the brain and the immune system can influence each other in a bidirectional loop. We have also seen that – like all perception – our experience of infection is a meeting of our brain's model of the world and the reality of physical matter. The nervous and immune systems are similar in nature, deeply intertwined and often mutually dependent. Finally, let's go one step further, and consider it all as a unified system.

The human body does not exist to house distinct organs and tissue-specific systems that neatly demarcate medical school modules and hospital departments. Often, of course, it is useful and necessary to categorize 'brain', 'mind' and 'immune system' as separate entities – for the sake of our limited human comprehension, if nothing else. But we should never confuse these conceptual boundaries with reality. Different organs and body systems have been created and welded together by powerful selection pressures over millennia in the pursuit of common goals. Survival, I think it's fair to say, is an important goal, and nested within this is a need to defend against threats. I believe that this need is best conceptualized – as it best reflects reality – as a supersystem: the 'defence system', which makes use of neurological and immune capabilities.

On the surface it seems that the nervous system is there to defend against the 'macro' threats – hungry lions, hot pans and hostile humans – whereas the immune system is equipped to fight 'micro' foes. Throughout human history, however, these threats regularly come in a package.[5] A lion's bite would often be accompanied by multitudes of microbes. Fighting off a dangerous pathogen also has consequences for your risk of being damaged by macro threats while you're too weak to defend yourself. Research clearly shows that when you are psychologically stressed – one study had its subjects try to solve tricky arithmetic tests and deliver a hastily prepared presentation to a panel of hostile judges – your immune system is activated.[6] You could say that your immune cells are listening to your thoughts. Conversely, when your immune system engages a pathogenic invader, your thoughts, mood and behaviours change, along with an increased sensitivity to pain. In the defence system, the neural and the immune are integrated, overlapping and enmeshed.

A curious idea that is only starting to be scientifically explored is that the defence system is a unified predicting machine. Perhaps it was Ivan Pavlov, the remarkable Russian psychologist, who laid the foundations for this discovery. Pavlov is best known for his work with dogs and bells. He was intrigued by the well-known observation that dogs salivated at the sight of tasty food. The food is called an 'unconditioned' stimulus as the dog doesn't need to learn anything to salivate when it sees the food. Pavlov then introduced a neutral stimulus, the sound of a ringing bell, whenever the dog was given food. It didn't take long before the ringing of the bell alone would induce salivation; it had become a 'conditioned' stimulus. This is called 'classical conditioning': prediction in practice. Almost a century later, in the 1970s, the American psychologist Robert Ader decided to do something conceptually daring and to see whether classical conditioning worked on the immune system.[7] His experiment was audaciously, heretically interdisciplinary. Ader took three groups of rats. He injected the rats in Group 1 with cyclophosphamide, an immunosuppressant that also makes one feel very sick, and at the same time fed them with sugar-flavoured water. Rats in Group 2 were also injected with cyclophosphamide but drank plain water. Group 3 were injected with a placebo and had plain water. Afterwards, all of the rats were injected with red blood cells from sheep: something that should trigger a strong immune response. After a few days, the rats were given the sugar-flavoured water to drink. Those rats that had earlier been given the horrible immuno-suppressant when they last drank the sugar-water became very averse to drinking it again. This finding is not hugely surprising; it's not uncommon to develop taste aversion to a food that once carried a bacterium that gave you gastroenteritis or to an alcoholic drink you once over-imbibed to the point of bringing it all back up again. But the curious thing that Ader found was that when these rats drank the sugar-flavoured water again – this time in the absence of cyclophosphamide – they still became immunosuppressed. Some even died. It seemed that the brain and the immune system were predicting together.

A predictive processing theory of the immune system was first fully articulated in 2021 by Karl Friston's team.[8] They first observed that the immune system works in a similar way to the brain: it is self-organizing and self-maintaining, and it has its own internal model of the environment. The immune system updates this model through sensing and acting upon changes in the microscopic envir-onment. Friston's research team argue that much in the same way as the brain combines bottom-up sensory data and top-down beliefs, the immune system carries out 'immunoceptive inference', which is when the immune system 'is seen as furnishing predictions of – and acting upon – sensory input, informing "beliefs" about whether an antigen belongs to the category of "self" or "non-self"'. Perhaps the number of a certain type of immune cell or receptor in the body at any one time represents an implicit belief that there are a certain number of that cell's intended targets in the environment. For example, an increase in the population of immune cells that target bacteria reflects a belief that the body is fighting a bacterial infection, not a viral one. But, crucially, this inference is not happening in isolation from the brain, which is constantly sensing the state of the immune system. Through the numerous bidirectional links between immune system and brain, this combined neuro-immune system is constantly predicting its environment and striving to reduce uncertainty. Friston's researchers argue that the human organism has 'a common generative model that the brain and immune system jointly optimise' in order to distinguish and respond to threats in the environment. This is all ultimately in the service of the minimization of uncertainty and the maintenance of body integrity. In short, we have a unified defence system that enables us to survive and thrive.

The Ministry of Defence

The complex, multifaceted structure of your defence system might make more sense when likened to a complex system with which we're much more familiar. Consider your whole being – brain and body – to be a kind of nation state. Tasked with the protection of this country is its Ministry of Defence. This is a complex, diverse, multi-departmental governmental organization specializing in distinguishing friend from foe and mounting an appropriate response to the latter. The enemy can come in many different forms: overtly hostile countries (other humans and large animals), rogue states and terrorist cells (pathogenic microorganisms), as well as undercover agents (bacteria and viruses that live and replicate within human cells). The Ministry of Defence has a diverse arsenal, but it needs to use it responsibly to avoid collateral damage or overspending the body's finely balanced energy budget. Different departments within the Ministry communicate through different channels. Some of those in the brain are attending a perpetual videoconference, ready to react with split-second immediacy to certain threats. Communication between neural and immune departments in the Ministry can also be relatively quick, via the telephone lines of nerves, or slower, through the emails and postal messages of cytokines and hormones. Crucially, the Ministry of Defence can learn and adapt to changing threats in an uncertain world, whether they be psychological, immunological or both. The metaphor also accommodates the predictive power of our defence system. To ensure the country's stability and survival in the most efficient way possible, the Ministry of Defence must anticipate threats and reduce surprise, continually adjusting its preparations in response to information coming from the outside world via surveillance and intelligence briefings.

This all describes a healthy system, but let me extend the metaphor to disease. A good defence force is one that protects the integrity of the nation without harming its own citizens. History, and our world today for that matter, is littered with cases of countries responding to real threats – wars, riots or terrorist attacks – by becoming a hypervigilant, oppressive police state. A medical translation of this hypervigilance is 'hypersensitivity'. Hypersensitivity affects all aspects of our defence system, resulting in the system attacking its own organism. Hypersensitivity can be immunological – from allergies and autoimmune disease – or psychological – from PTSD to phobias. The concept of a hypersensitive defence system can also help explain why diseases of the immune system and the mind so often occur simultaneously. If one part of the defence system is unbalanced, it's likely that the other elements will be, too. It is well established that a high number of infections in early life increases the risk of mental health conditions later in life and, conversely, early-life psychological stress can lead to inflammation.[9] Their interconnectedness is another clue that, in both cases, a unified defence system is being primed to hypersensitivity, the consequences affecting mind and body.

The modern, developed world is experiencing an ever-growing crisis of both chronic immunological diseases and mental health conditions. Perhaps this is not a coincidence, but is instead a pandemic of unbalanced defence systems. This is partially because the bodies and minds we have inherited from our ancestors were honed to survive in a very different environment. Until the health and sanitation revolutions of the twentieth century, humans in all strata of society lived in differing layers of filth, with no access to antibiotics or vaccines. This resulted in a high infectious mortality; those who survived tended to have more aggressive immune systems, prone to err on the side of inflammation. It was also a world of constant inter-human violence and regular exposure to predatory animals. Psychologically, this would have favoured a tendency towards both anxiety about others and also depression-like sickness behaviour during infection.[10] Across human history, from generation to generation, there has thus been a leaning towards an 'inflammatory bias'. And selection for this inflammatory bias has not always been gradual. Between 1346 and 1350, the Black Death wiped out between one third and one half of the populations of Europe and Asia. A fascinating study, published in October 2022, analysed the genetic material of medieval Londoners who lived before, during and after this deadly plague.[11] The remnant of the population that survived had much higher frequencies of genes that confer immunity to *Yersinia pestis*, the bacterium responsible for the Black Death. But this came at a cost: some of these genetic variants are risk factors for the autoimmune diseases rheumatoid arthritis, lupus and Crohn's disease. In the same way that the military culture of a society is inherited from generation to generation – ranging from open and tolerant to hostile and defensive – the culture of our defence system is passed down to us through DNA.

One positive consequence for our ancestors living in filth was that, from an early age, their immune systems were vigorously trained to differentiate friends (food and 'good bacteria') from pathogenic foes. In today's sanitized, antibiotic society we are less likely to have this opportunity to help our immune system develop tolerance, refining its predictions about what is 'self' and what is 'non-self', what is safe and what is a threat. This is the core of the 'hygiene hypothesis', which is a widely accepted theory to explain the staggering increase in rates of allergy and autoimmunity in the modern world. This does not mean that we should actively seek to be infected with pathogens, running around crowded train stations in the hope of catching Covid, but instead suggests that we should seek to harbour in and around us a rich microbial world (more on that in the next chapter). So, as modern humans, have the worst of both worlds: the inflammatory bias of our ancestors without the environmental tools to develop immune tolerance. A parallel argument can be made for mental health and stress. Our ancestors would frequently be exposed to mortal peril and have to mount stress responses that resulted in memories of the threats, adaptive behaviours and immune responses. The threats facing most people in the modern world tend not to be of extreme, existential violence, but are a slow-but-steady drip-feed of stress: busy corporate life, unemployment, mortgages, insomnia, constant comparison on social media, to name just a few. These often activate an inappropriate immune response, the long-term consequence being chronic inflammation. The speed of technological development over the past century has, in evolutionary terms, transported humanity to a foreign, unnatural world. No wonder it is so easy for our defence systems to become unbalanced, miscalibrated and – in many cases – hypersensitive.

The concepts in this chapter are radical, particularly in a world that still unthinkingly treats mind and body as unrelated. But I hope that they provide you with the tools to explore a new view of how humans work, how they go wrong and how they can be healed. The concept of a unified defence system helps explain how the mind and the immune system – two seemingly unrelated entities – work together in sickness and in health. A disruption to one part of a system can ripple across the whole supersystem. A predictive processing model of perception provides another perspective with which we can heal diseases of systems from both top-down and bottom-up, changing both model and matter to minimize uncertainty.

In these first few chapters we have seen that, contrary to many of the orthodoxies of the medical world, your immune system and your mind are enmeshed on an anatomical, physiological, clinical and conceptual level. But there is one final layer to explore in this first part of the book. I hope you've brought your microscope.

5

Your Mind on Microbes

How the trillions of guests in your gut influence your choices, mood and behaviour

> *Disease usually results from inconclusive negotiations for symbiosis, an overstepping of the line by one side or the other, a biological misinterpretation of borders.*
>
> LEWIS THOMAS,
> *THE LIVES OF A CELL*

Meet Tom, Jerry and Terry. Tom is a ham-fisted house cat who is unfailingly unsuccessful in his quest to catch Jerry, a cute brown mouse that shares Tom's home. Terry, less familiar to most, is a *Toxoplasma gondii* parasite. This sesame-seed-shaped microorganism is only able to sexually reproduce in the feline gut, so Terry begins his existence somewhere within Tom's gastrointestinal tract. This marks the beginning of a remarkable life. The infant Terry is released through Tom's faeces into the outside world. A growing *Toxoplasma* parasite needs other animals – 'intermediate hosts' – for it to fully develop before returning to a cat's gut, and as Jerry explores and nibbles at his environment while Tom is out of the house, he unwittingly ingests Terry. What happens next is astounding. The parasitic

Terry invades Jerry's brain, causes inflammation and – through pathways as yet undiscovered by scientists – changes Jerry's behaviour. Like a human after a few glasses of bubbly, Jerry begins to let down his guard. He gradually grows in confidence and begins to lose his fear of predators and his aversion to feline scents.[1] When Tom returns to the house after prowling around the neighbourhood, he finds that the previously cunning and skittish Jerry is now dangerously nonchalant. Dinner is served, and the fully-grown Terry can return home.

This is perhaps one of the clearest cases of a microorganism as puppet-master, bending the will of its host for its benefit. There are, thankfully, few examples of individual pathogens directly manipulating human minds to aid their spread; the aggression-causing rabies virus is a rare example. But a growing body of evidence tells a compelling story of how our mind is moulded and manipulated by the trillions of microbes that live within us. We will find out in this chapter that, unlike Jerry and Terry, our relationship with the microbes within us – collectively known as the microbiome – is mostly mutually beneficial. Our mind needs microbes. Not only this, but we will see that this population of microorganisms within us can be considered, alongside the mind and the immune system, as part of our body's defence supersystem. We begin our journey over a century ago, among whispers in the scientific community that the great Élie Metchnikoff was starting to lose it ...

The Bulgarian solution

It was 1908, and the sixty-three-year-old Ukrainian scientist Élie Metchnikoff had just been awarded the Nobel Prize in Medicine for his pioneering discoveries in the field of immunology over the previous few decades. But in his later career – fitting the mould of the stereotypical 'mad professor' – Metchnikoff had started to espouse some rather outlandish theories. To the embarrassment of his colleagues at the Pasteur Institute in Paris, and to the amusement of scientists across Europe, Metchnikoff was beginning to become increasingly evangelical about what he believed to be the elixir of health and longevity: yoghurt.

Élie Metchnikoff was born in 1845 near what is now Kharkiv, Ukraine, and was then the Imperial Russian city of Kharkov. A prodigious intellect, he completed the four-year Natural Sciences degree at Kharkov Imperial University in just two years. His childhood nickname of 'Quicksilver' reflected his adventurous spirit and restless mind. After a few years travelling between German universities and studying the comparative anatomy of various crustaceans and worms found across Northern Europe, he became a lecturer at Odessa University. Aged twenty-two, he was younger than most of his students. A decade later Louis Pasteur, one of the fathers of microbiology, invited him to work at his institute in Paris. It was here that Metchnikoff would make his Nobel-winning discovery. He found that human immune cells could remove and destroy pathogenic bacteria by engulfing them, a process known as phagocytosis.[2]

In his later years, Metchnikoff's interests focused on the ageing process, specifically how influencing the immune system and the body's response to bacteria could prolong the human lifespan. His attention was turned to his native Slavic lands, where a number of poor, rural areas seemed to yield a disproportionate number of centenarians. He honed in on a group of long-lived Bulgarian peasants, and theorized that the source of their vigour and longevity was from their daily ingestion of soured milk, called 'yahourth'.

Metchnikoff had lived through – and indeed had been part of – the golden age of microbiology, in which the 'germ theory' of disease had utterly revolutionized medicine. By the late nineteenth century it was clear that many diseases were caused by microscopic pathogens that infected humans, caused sickness and spread between hosts. Naturally, it was widely believed that all germs were bad for the body. A logical continuation of this belief was all gut microbes were bad, and their breakdown of food was really a process of rotting, releasing poisonous, disease-causing chemicals into the body; in the late nineteenth century the influential French doctor Charles Bouchard even called the colon 'a laboratory of poisons'.[3] Metchnikoff, however, found that giving mice the Bulgarian yoghurt – milk that had been fermented by lactic-acid bacteria – enabled them to thrive and produce more progeny. He practised what he preached, and took the yoghurt daily for the rest of his life, extolling its positive effects on his health. Scientists at the time assumed that these positive effects must stem from lactic acid having some kind of antiseptic property, halting the proliferation of microbes in the gut. Metchnikoff and his colleagues, however, found that giving lactic acid on its own did not help; the mice actually seemed to be benefiting from the bacteria in the yoghurt, which they isolated as *Lactobacillus bulgaricus*. From his findings, Metchnikoff summarized a revolutionary theory in his 1907 book *The Prolongation of Life: Optimistic Studies:* 'The general belief is that microbes are all harmful. This belief, however, is erroneous. There are many useful microbes, among which the lactic bacilli have an honourable place.'[4] In this book he also articulates a vision of the future of medicine: 'The dependence of the intestinal microbes on food makes it possible to adopt measures to modify the flora in our bodies and to replace the harmful microbes by useful microbes.'

This vision, unfortunately, was about a century ahead of its time. For one thing, the technologies that could properly analyse the composition of microbes in the gut, let alone their effects on the body, were many decades away. Perhaps more importantly, the world was not ready for the idea that microorganisms could play a positive role in human health. The relatively new germ theory of disease had led to the discovery that microbes were responsible for the great killers of the age: cholera, malaria, smallpox, scarlet fever, syphilis and typhus. Society's new aseptic attitude had no room for the concept of 'good bacteria'. My colleagues who study the potential benefits of psychedelics and cannabis-derived molecules in treating mental illness often bemoan Nixon's 'war on drugs' in the 1970s, which terminated a number of rich seams of medical research and held the field back for decades. Similarly, Élie Metchnikoff's theory was a victim of the early twentieth century's 'war on bugs'. And it was only going to escalate. A decade after he outlined his theory, an H1N1 influenza (commonly known as Spanish flu) pandemic would kill more people than the First World War. A decade after that, the British microbiologist Alexander Fleming would return to his lab in St Mary's Hospital, London, to find one of his bacterial plates contaminated with a fungus from the lab on the floor below. Most of the bacteria on the plate had been killed by the mould, which turned out to be *Penicillium*. The discovery of antibiotics was one of the greatest achievements in human history; the antibiotic age has saved countless lives and relieved an immeasurable amount of human suffering. But a side effect was that it reinforced the idea that microbes are always the enemy. In light of these discoveries, Metchnikoff's research on yoghurt-guzzling Bulgarian peasants seemed at best insignificant, at worst grossly mistaken. We were entering a brave, bug-free world.

Of mice and microbes

The mid-twentieth century saw a breakthrough in animal research: the germ-free mouse. These are mice bred in aseptic incubators that shut out any environmental microbes. Germ-free mice – lacking any microbes in or on them – presented an opportunity for researchers to study mice without any contaminants. They are biological blank slates: you can study the effects of an infectious pathogen without the confounding presence of pre-existing microbes. But towards the end of the century, it had become very clear that these squeaky-clean rodents were not paragons of health. Research throughout the 1990s had partially vindicated Metchnikoff's theory – gut microbes appeared to positively influence gut health, playing a part in providing the host with nutrients and fine-tuning its developing immune system. But by the turn of the twenty-first century, researchers began to notice something extremely odd: germ-free mice *behaved* differently.

In 2004, a group of Japanese scientists carried out a groundbreaking study. They compared the brains of germ-free mice to those raised with gut microbes.[5] They found that germ-free mice are more stressed: they produced abnormally high levels of the stress hormone corticosterone (a rodent's equivalent of cortisol) compared with normal mice. The scientists discovered that they could resolve this by feeding the mice the *probiotic* (health-conferring) bacterium *Bifidobacteria infantis*. Conversely, when the germ-free mice were given the pathogenic bacteria *Escheria coli*, they exhibited an even higher hormonal stress response. Germ-free mice also had lower levels of brain-derived neurotrophic factor (BDNF), a protein critical for learning and memory formation. Intriguingly, the scientists found that treatment with probiotic bacteria only reduced stress if given in the first few weeks of life, suggesting an important early period in which gut microbes help sculpt the brain's stress circuitry.

Why labour over the brains of lab rats? Understanding how gut microbes might influence brain health is a field in its infancy and animal models are crucial for establishing mechanisms before exploring whether these effects also occur in humans. We are at an exciting juncture in scientific history in which we can study these mechanisms in detail, and are starting to see how they are relevant to us. To find out more, I spoke to the godfather of the gut-microbe-brain connection, Professor John Cryan of University College Cork, Ireland. 'I was a hard-core neurobiologist and neuropharmacologist, and I think when colleagues saw me going back to Ireland and playing around with microbes, they probably thought I'd gone feral,' laughs Cryan when I speak to him via video link. In the background,

I can see his desk, nestled between a tome-laden bookshelf and an eclectic mix of contemporary paintings and porcelain models of the brain. When he was younger, Cryan would never have imagined that his career would revolve around gut bugs. After completing his PhD in pharmacology in his native Ireland, he pursued a number of postdoctoral fellowships in the USA and Australia, looking into the neurobiology of stress. In 2005, he returned to Ireland to work on how stress affects the brain and behaviour of rodents. But it just so happened that some of his colleagues were starting to explore the recently discovered world of the microbiome – that ecosystem of microorganisms that live in and on us. In 2007 Cryan and his team decided to see whether psychological stress has any effect on the largest microscopic community: the gut microbiome. 'We found that the composition of microbes in the gut of animals who had been stressed in early life was different.[6] Back then, we knew that stress in early life affected the immune system and the health of the gut, but this was the first time that we could see the effects on the gut microbiome as well. It is clear that psychological stress is a whole-body syndrome.'

Cryan and his colleagues were aware of the Japanese team's work on germ-free mice and – alongside groups in Sweden and Canada – carried out studies that confirmed that a healthy brain does not develop in the absence of gut microbes.[7] 'The final part of the puzzle was that if an abnormal microbiome causes an excessive stress response, we should be able to change the microbiome in a positive way and dampen down that stress response,' Cryan tells me. They did this by following in the footsteps of Metchnikoff, finding that feeding mice the probiotic *Lactobacillus rhamnosus* reduced the behavioural and biological responses to stressors.[8] 'But we went one step further and showed that not only did *Lactobacillus* affect brain, behaviour and stress, but if we cut the vagus nerve, all of the effects were gone.'

We came across the vagus in Chapter 2: a long nerve that, on its meandering journey down from the brainstem, communicates with most of our internal organs, including the gut. Cryan's team found that the vagus also communicates microbial signals from the gut to the brain: 'As I like to remind people, what happens in vagus, doesn't stay in vagus!' It seems that molecules produced by gut bacteria can be detected by the brain via the vagus nerve, with potentially profound consequences for the brain and mental health. Since these seminal findings, between 2007 and 2011, Cryan's lab has been primarily focused on understanding the numerous ways in which microbes affect the brain and behaviour.

Further studies from Cryan's lab revealed that germ-free mice had a messed-up relationship with anxiety. One study found that they had reduced anxiety-like behaviour compared with normal mice, which, on the face of it, sounds like a rather good thing. But it soon became clear that their fear-related recall – their ability to recognize threatening stimuli – was impaired.[9] When regular mice were given an electric shock directly after a tone, they learned to associate the tone with the shock, to the point that when they heard only the tone they froze, anticipating the shock to come. This is a case of Pavlov's classical conditioning. Germ-free mice, however, didn't learn to associate the tone with pain, and continued as normal. If these mice, however, were given gut bacteria, they showed a normal, appropriate anxiety response. But, like their earlier study exploring excessive stress in germ-free mice, these effects were only seen if the gut bacteria were given early on in life, suggesting a critical developmental window. Later studies found a potential source of these odd behaviours: the amygdala, two almond-shaped clusters of nerves located deep within the brain (*amygdala* being the Greek for 'almond'), each located a few inches directly behind each eye. These play key roles in emotional response to stimuli, including 'fight or flight'. The amygdala of germ-free mice, however, is unusually large and has multiple abnormalities in its structure.[10]

These behavioural studies have been complemented by research demonstrating abnormalities in the structure of the brain. One remarkable finding, from a 2016 study, was that microbe-deficient mice have very strange prefrontal cortices.[11] The prefrontal cortex is an advanced, complex area of the mammalian brain that – as its name suggests – sits right at the front of the organ. Part of the brain's frontal lobe, it plays key roles in the development of personality, self-control, and the integration of thoughts and feelings in the context of long-term goals. Such is its importance to the working of the mind – including decision-making, planning, personality and behavioural control – that a dysfunctional prefrontal cortex is implicated in most mental health conditions. The nerve fibres in the prefrontal cortex (along with most of the rest of the nervous system) are wrapped in an insulating material called myelin. Normal levels of myelin are critical for a well-wired brain. But when the researchers looked down electron microscopes at the prefrontal cortices of germ-free mice, they saw something that very rarely occurs in nature: too much myelination. This odd brain development could be reversed if the mice were colonized with normal microbes.

Alongside germ-free mice, numerous other research techniques are helping us understand the gut microbiome and its effects on behaviour. This includes removing the normal gut microbiomes of animals by giving them long-term antibiotic treatment, resulting in dramatic brain and behaviour changes similar to those seen in germ-free mice.[12] There is also the 'microbiome transplant', which is exactly what it sounds like: inserting what is essentially freeze-dried poo into the recipient's gut, either given orally or rectally. If we can observe changes in behaviour after the introduction of a faecal transplant, it is suggestive that it is at least partially driven by the microbes inside it. Microbiome transplants might one day provide an option for mental health treatment for humans – either in the form of enemas or 'crapsules'. One curious 2011 study found that these transplants help normalize the social lives of germ-free mice. Alongside abnormal stress and anxiety responses, it is well known that germ-free mice tend to be loners, exhibiting significantly reduced social engagement and a lack of willingness to explore their environment. This study, however, found that if you transplant the microbiome of a healthy mouse into its germ-free cousin, the germ-free mouse springs to life and begins exploring its environs.[13] A few years after that study, John Cryan's team transplanted the microbiota of humans with depression into rats, which brought about depressive behavioural symptoms in the rodents.[14] This was not just a microbiome transplant: it was a cross-species depression transplant.

It appears that – in mice, at least – gut bacteria are not only good for the body, they are necessary for the development of a healthy brain. This goes some way towards vindicating Élie Metchnikoff's much-derided theories. But now we need to see how this translates to the human organism.

You are a community

You are not alone. Whether you like it or not, you are teeming with microbial friends. Tens of trillions of them. You have at least as many microbes living with you as there are cells in your body, and probably more.* But when it comes down to genetic material, prepare to be completely humbled: 99 per cent of the genetic material you carry around with you is microbial. You are a world frothing with life, a planet housing myriad ecosystems and biomes. I also don't use the word 'friends' glibly; most of the organisms that live in or on you are beneficial 'mutualistic' or 'probiotic' microbes. Even those that don't seem to confer a direct benefit – 'commensals'

---

* It's commonly said only 10 per cent of the cells in our body are human, but this has recently been revised to 50 per cent. That's still staggering. Note that these are estimates and the jury is still out – it's like trying to guess the number of grains of sand in a desert.

coming along for the ride – reduce the chances of bad microbes (pathogens) infecting you, simply by outcompeting them for space. Almost every surface of the body is covered in microorganisms, but the vast majority of your microbial companions reside along the 9 metres of your gastrointestinal tract, mostly in the colon. Here they form a thick layer, hundreds of grams in weight. It remains relatively stable in composition despite being constantly on the move; roughly 60 per cent of your faecal matter is microbial. I say 'relatively' stable as it can be changed to varying degrees by how we sample our environment – namely, what we eat and drink. The gut microbiome is not simply a boarding house for bacteria, however: it is our secret organ. It is a fermentation factory that breaks down dietary fibre and other long-chain carbohydrates – a delicacy for bacteria, which cannot be broken down by our own cells. The molecular by-products of these microbial feasts include short-chain fatty acids (SCFAs), which circulate around the body like hormones, delivering diverse effects for your immunity and metabolism.[15] You are effectively out-sourcing the production of essential molecules to microbes. But the gut microbiome is not just a fancy food processor: it is also an immunological organ. Your microbiota can directly defend against pathogens. This is most clearly seen in the absence of a healthy microbiome: a long course of antibiotics greatly increases the risk of a *Clostridium difficile* infection in the gut. As the gastrointestinal tract is essentially a long tube directly exposed to contents from the outside world, the gut microbiome is a sensory interface between the external and the internal. If our body is a country, the gut is a heavily defended international border: we have more immune cells in and around our gut than we do in our blood and bone marrow put together. This border force needs to be adept at differentiating the *threatening* from the *safe*: from tolerating food to attacking pathogenic microbes. This is where the gut microbiome comes in. Throughout life, but particularly in early life, a diverse microbiome teaches our immune cells to create inflammation in response to pathogens, but to suppress it in the presence of friends: 'immune tolerance'.[16] The microbiome is perhaps the immune system's largest training academy, its West Point or Sandhurst.

The gut microbiome is a multifaceted organ made up of cells that are not your own, but which you could not live without. This bizarre fact only begins to make sense when we take a wider perspective. Humans are part of a microbial world, and we came late to the party. There has been no point in history when the human body has not been bathed in microbes, and we have shaped each other over millennia, with mutual benefit. It is easy to view natural history as a dog-eat-dog fight for domination, but the arc of nature bends towards cooperation. The deal we have made with microbes is this: we give them environments in which they can live and thrive, and they help us to live and thrive. They get a contained environment and free transport, while we can now digest and benefit from the nutrients in fibrous foods, as well as being better positioned to fight pathogens. This is a bilateral agreement that puts the United Nations to shame. The overarching takeaway from this is that we need to keep our side of the deal: a thriving environment for microbes is good for us.

Mind-altering microbes

The gut is our largest reservoir of microbes and it is our largest immune organ; this interface is clearly vital to our health. But an ever-growing body of research is attempting to answer a curious question: could microbes be hacking our mind?

There are few clear-cut examples of pathogens manipulating human behaviour for their own survival. At the beginning of this chapter we met Terry the *Toxoplasma gondii* parasite, and saw how he lowered the inhibitions of Jerry the mouse, his intermediate host, so that he could find his way back into the feline intestines of Tom. There are theories positing that humans who are infected with toxoplasmosis also change their behaviour to prefer living in close proximity to cats, with a resulting neuroinflammation that can tip into psychosis: one possible origin for the negative stereotype of the 'crazy cat lady'.[17] The rabies virus is spread between hosts through broken skin and saliva, so in order to spread it makes its hosts extremely aggressive and, for want of a better word, bitey. But new research is suggesting that many of our mundane behaviours – particularly our food preferences – can be hijacked by the needs and wants of our gut microbes. An important study, published in 2022 by researchers at the University of Pittsburgh, found that if you transplant the microbiomes of different types of wild rodents (herbivorous, carnivorous or omnivorous) into germ-free mice, you can determine their type of foraging behaviour and what they choose to eat.[18] Germ-free mice also have an insatiable appetite for sugar, which is dampened following the introduction of a microbiome.[19] These findings have been experimentally seen in other animals, and while conclusive studies on humans are lacking, it is plausible that microbes have some say in setting the menu.[20] This might explain why our cravings for certain foods change following significant dietary alterations. Different microbes have different tastes – *Bacteroidetes* love fat, *Prevotella* have a penchant for carbs and *Bifidobacteria* are addicted to fibre – so the makeup of our gut microbiome probably both reflects and influences our food choices. An addiction to fast food might partly be down to fat- and sugar-loving microbes in your gut making persistent delivery requests. As behaviour – including willpower – is partly microbial, if you want to change your diet to improve your health, perhaps altering the microbial environment is better than simply relying on willpower.

It is becoming clear, however, that the gut microbiome plays a much larger role in our mood, thoughts and behaviour than just dictating our dietary choices. Humans have always had an instinctive 'gut feeling' that our mind and gut are interlinked. This is seen in language: 'go with your gut', 'I'm feeling gutted', 'I've got butterflies in my stomach'. From a conceptual point of view, it makes sense that the brain should be linked to the gut. As we have seen, the brain is a predictive machine trying to balance the body's energy demands, and our brain and immune system are part of a defence system trying to differentiate friend from foe. As the lining of the gut is our body's largest interface with the outside environment, gut-brain communication is crucial in establishing balance and ensuring survival. And as our gut microbiome is a key element of our gut, an organ in and of itself, it should be no surprise that there are plenty of biological roads linking this microscopic community with the brain.

There are four primary pathways between the gut microbiome and the brain: neural (relating to the nervous system), endocrine, metabolic and immune. Our gut is sometimes termed our 'second brain', as it has its own nervous system (the 'enteric nervous system') that contains more nerves than the human spinal cord. Its most well-known role is unconsciously and rhythmically squeezing food down the pipe of the intestines, but – as we've seen – the gut is also directly linked to the brain via the vagus nerve. The vagus has long 'feet' that dig into the lining of the gut, and in 2015 it was discovered that these feet contain 'neuropod cells' that taste the environment of the gut and directly pass signals up to the brain.[21] This is already having astounding implications for our knowledge of how we can sense nutrients outside of the mouth, including an explanation for why we tend to prefer natural sugars to artificial sweeteners, even when we can't tell them apart by taste.[22] It is becoming evident that gut microbes can also relay molecular signals up to the brain via these cells,[23] adding strength to the idea that there is a constant, mostly subconscious, dialogue between gut and brain via the vagus that shapes our decisions.

The microbiome-nervous system communication is clearly bi-directional, as was neatly shown by two studies published in October 2022. The studies revealed that pain-sensing neurons in the gut can not only detect molecules produced by gut bacteria, but can also produce molecules that regulate the types of bacteria around them, shaping the microbiome.[24] Microbe-derived molecules are also capable of stimulating endocrine (hormone) cells in the gut lining – called enteroendocrine cells – that produce various different types of hormones. These hormones interact with the vagus nerve and brain to influence hunger and eating preferences. Microbe-produced molecules, particularly short-chain fatty acids (SCFAs), can also directly communicate with the brain.[25] What's more, gut microbes are capable of synthesizing the precursor molecules of neurotransmitters and neuromodulators that are critical to brain function, such as serotonin, dopamine, norepinephrine and gamma-aminobutyric acid (GABA).[26] And, of course, the intimate relationship between the gut microbiome and the systemic immune system means that ripples from this relationship can reach the brain in the form of cytokines or immune cells. This is the briefest of summaries of the beautiful, complex dance that is the 'gut microbiota-brain axis'. The key message is that we now know there are many ways in which our gut microbiota can communicate with our mind. And this dialogue affects every stage of life.

Friends for life

In microbial terms, you are born as a blank slate of flesh, thrust into a world crawling with microbes.[27] Your noughth birthday present is a bacterial blanket from your mother's birth canal. Her vaginal microbiome has changed considerably during pregnancy, multiplying colonies of beneficial *Lactobacillus*, bolstering your microbial starter kit. It appears that those born by Caesarean section (C-section) are more prone to allergies later on in life,[28] but it is hard to tell whether it is the method of delivery itself that causes this, or the fact that those born by C-section are more likely to have had in-utero problems or mothers with pre-existing health issues. Still, one argument in favour of the delivery hypothesis is that those born by C-section are more likely to have an unbalanced microbiome that is inadequately diverse and that can harbour opportunistic pathogens.[29] This persists for at least the first year of life and the effects may last longer. This early dysbiosis (an unbalanced microbiome) may result in a poor learning experience for the young immune cells in the gut's military academy. When stimulated, the immune systems of those who have been born by C-section produce inappropriately high inflammatory markers compared with vaginally-delivered babies.[30] This is an immune system that has not developed tolerance; one that is inappropriately skewed towards inflammation. Remarkably, but perhaps not surprisingly, the mental part of the defence system also becomes hypervigilant. A 2022 study found that adults who were born via C-section were more likely to exhibit a stronger psychological stress response, as well as a greater immune response, to both short-term stressors and during a longer-term period of stress, such as exam season.[31] The last thing I want to do is discourage people from C-sections – they can be life-saving. Nor do I want to cast a fatalistic shadow over everyone delivered this way. There are many other environmental influences on the microbiome, and the microbiome is an environment itself that can be changed for the better.

In fact, a study that analysed over 12,000 stool samples from around 1,000 infants over the course of their early childhood found that the most significant influence on the gut microbiome is breastfeeding.[32] Human breast milk contains the highest complexity of sugars of any mammal. What is truly extraordinary is that the human intestines are incapable of breaking down long sugars (called oligosaccharides) and digesting them into the molecules that are vital for body and brain health. Not so for our gut microbiome. This may at first seem utterly bizarre, but a significant portion of breast milk is not food for the baby, but for its gut bacteria. The nourishing of bacterial life is necessary for the nourishing of human life; we can't separate the two. It is also apparent that some bacteria – potentially even from the mother's gut – are found in breast milk.[33] Breastfeeding is strongly associated with positive health outcomes, including mental health. It is associated with improved cognition in children and teenagers, and while there may be other factors at play, this finding remains even when numerous socioeconomic factors are taken into account.[34]

Research in this area is truly in its infancy (no pun intended), but there are now scientists exploring the relationship between the gut microbiome and the perinatal period in animal models. One of these, a colleague of mine called Daniel Radford-Smith, is a researcher at Oxford University's Department of Pharmacology. Along with others in his lab, he has found that improving the microbiome of mice mothers has profoundly positive effects on their offspring.[35] Daniel has discovered that probiotic supplements given to mouse mums were sufficient to increase the levels of microbially-derived short-chain fatty acids (SCFAs) in their milk during nursing, whereas those just on a high-fat diet had decreased SCFA levels. These SCFAs support brain growth and neural plasticity during key neurodevelopmental windows in offspring, which likely promote resilience in the mouse pup that lasts into adulthood. Levels of brain lactate – a molecule associated with the brain's energy supply – were also increased. There were similar findings with 'prebiotics', which are not bacteria but a source of food for existing gut microbes.[36] While these studies are yet to be carried out in humans, Daniel's research certainly provides hope for future dietary plans or probiotic interventions to help parents give their offspring the best immunological and psychological start to life. In fact, in a recent study on which Daniel and I collaborated, findings from a huge database of British individuals found that those with high blood levels of the molecules involved in the generation of energy in the body – including lactate – were less likely to develop depression later on in life.[37] Given that Daniel's team has already found an association between probiotic use in early life, lactate levels and stress resilience in mice, it might be the case that a healthy microbiome provides us with the molecules that enable us to thrive and maintain resilience in the face of stressors.

After weaning, it appears that the next few years of life are particularly crucial for gut microbiome development. There seems to be an initial developmental phase between three and fourteen months, as breastfeeding establishes a microbiome full of healthy bacteria but one that is not hugely diverse, before a transitionary period from fifteen to thirty months, in which the microbiome diversifies and then stabilizes. Perhaps an infant's frenzied oral exploration – from eating dirt to licking the cat – is a way of sampling their environment to diversify their gut microbiome, train their developing immune system and, ultimately, fine-tune their body's defence system. By the time you are three years old, you have unintentionally calibrated your body's defences. Microbes are our teachers, and they are integral to the development of a healthy immune system and a healthy mind.

Our microbes may also be integral in making friendships. As you'll recall, when John Cryan examined the behaviour of germ-free mice, not only did they have impaired stress tolerance and impaired anxiety behaviours, they also appeared to exhibit striking social deficits. 'Usually mice are very sociable and rather fickle, and they will prefer to spend time with a new playmate than an old one. But germ-free mice were much less adventurous,' Cryan told me. Germ-free mice are less sociable than normal mice and exhibit more repetitive self-grooming behaviours.[38] This emerging data may have implications for conditions associated with social behaviour, from autism spectrum disorder (ASD) to social anxiety. ASD, as its name suggests, is a wide and diverse spectrum, and it is dangerous to pigeonhole it as a single entity or excessively extrapolate from findings in animal studies. It is also important that any research on ASD recognizes the desire of many people to view autism as part of the spectrum of human thinking – neurodiversity. Interestingly, there has been a large increase in the prevalence of ASD in recent decades that is not fully explained by increase in diagnosis. In 2019 a team at the California Institute of Technology suggested that one cause may lie in the gut microbiome. They were able to bring about repetitive behaviours and reduced sociability in young mice by transplanting the microbiomes of humans with ASD.[39] The microbes in the intestines of these mice produced insufficient quantities of 5-aminovaleric acid (5AV) and taurine, molecules crucial for the synthesis of GABA, a neurotransmitter implicated in ASD. Other research showed that mice with ASD-like behaviours caused by induced maternal inflammation have their symptoms dramatically reduced if they are given healthy gut bacteria.[40] But mice are not men, and there isn't yet compelling evidence that gut dysbiosis causes or contributes to autism. It is, however, widely agreed that humans with ASD tend to have less diverse gut microbiomes than the rest of the population. What is debated is whether this is independent of food intake, or whether it is simply the result of more selective palates.[41]

But even if the data in humans is currently inconclusive, it appears that animals seem dependent on microbes for establishing normal social interactions. Why is this? One curious hypothesis, which John Cryan is exploring, is that microbes might make their hosts more social in order to maximize their ability to travel and reproduce.[42] If we are huge microbial ships, what's not to say that microbes could be at the helm? Perhaps these forces even influence kissing behaviours: it certainly can't do much harm to sample another microbial world before the establishment of a longer-lasting connection. An intimate ten-second kiss results in the transfer of around 80 million bacteria.[43] This is not to say that microbes intentionally manipulate mammals, like Remy the rat from the 2007 animated film *Ratatouille*. Instead, these behavioural effects may be the results of a gradual, trial-by-error tuning of microbe–animal interactions to maximize survival for both parties. One can get carried away with speculation, but Cryan for one believes that our gut microbes are 'friends with social benefits'.

Another intriguing role of the gut microbiome can be seen at the other end of the lifespan. A century ago, Élie Metchnikoff's obsession with Bulgarian centenarians and their love of probiotic yoghurt was driven by a desire to rein back the ravages of ageing and extend the human lifespan. John Cryan and his team at University College Cork took up the mantle: 'In 2013, some of my colleagues found that in an elderly population they had a lack of diversity in their microbiome, and the lack of diversity correlated with frailty and poor health outcomes.'[44] They had found that this was probably driven by a bland, 'beige' diet. Cryan wanted to find out how this worked on a mechanistic level, so his team explored the brains and bodies of ageing mice. Alongside the expected findings of cognitive impairment, the mice also had increased gut-barrier permeability, inflammation and an altered gut microbiome.[45] The next question was whether they could intervene by targeting the microbiome. 'For obvious reasons, I decided to focus on middle age,' he chuckles, 'so we gave middle-aged mice a diet high in inulin, which is a prebiotic found in vegetables such as leeks and chicory. We saw that even in middle age, the brains of mice showed some signs of inflammation, but we were able to slow this down by giving them inulin.[46] The take-home message is that if you are thinking of having a midlife crisis, instead of turning to motorbikes, look to leeks instead!'

By the time this study was published, in 2020, there was a lot of circumstantial evidence, but it still wasn't conclusive that diet could affect the brain independent of the microbiome. Cryan's next experiment was designed to answer the question, 'If we believe that the secret to healthy brain ageing is through the microbiome, then we should be able to take the microbiome from young animals and give it to old animals and rejuvenate the brain.' And it worked: by transplanting the microbiomes of younger mice into elderly rodents, his team were able to reverse age-related immune changes in the animals' brains and improve cognition and ageing-related anxiety.[47] This seems far-fetched, but since its publication in 2021 two independent teams have replicated its findings. It seems that faecal transplants rejuvenate the ageing brain. But, with a measure of caution, Cryan caveated: 'Am I advocating that the middle-aged go out and steal the poo of their juniors? Not quite yet. But this is part of accumulating evidence that if you want to mind your mind, you need to mind your microbes.' Maybe in the not-too-distant future there will be a market for the young to sell their faecal matter to an older generation in search of an elixir of life, and a way for Gen Z students to tap into the Baby Boomer treasure chest to pay their college fees. But while the animal studies are compelling, we need to wait for the conclusive evidence from human trials.

John Cryan is starting to feel redeemed for his investment in what has been seen as a controversial field. 'When I was asked to give a talk at an Alzheimer's conference in the USA in 2012, the audience could not have been less interested. Few people thought that the microbiome could play any role in neurodegenerative diseases. Now, every conference on neuroscience and psychiatry has some focus on the microbiome. I now feel a bit less like John the Baptist in the wilderness. We are gaining disciples.'

Inward awe

The human body has been bathing in microorganisms for its entire existence – in fact, there has never been a time in which the human brain has not been bombarded by microbial signals. We have developed a complex biological language that brings mutual benefit to host and guest. The gut microbiome provides vital nutrients to the body, protects against infection and trains the immune system. I think that we can firmly consider it as part of your body's unified defence system; an additional layer on top of your mind and immune system. We have already seen that stress in one part of the defence system often affects the other, whether it is an overactive immune system affecting the mind, or mental distress unbalancing the immune system. It's clear – in animal models, at least – that an absent or unbalanced micro-biome (dysbiosis) has profound effects on the mind and the immune system. It also appears that stress to either of these systems changes the makeup of the microbiome. As we will explore, we are now starting to see these effects in humans, which will have profound implications for mental illness. I ask John Cryan what he thought about visualizing these seemingly separate components as one unified system. His answer is characteristically imaginative: 'I certainly think that we can consider them the same system – this is something I've given some thought to. As I'm Irish, I like to picture them as a clover: mind, immune system and microbiome as three parts of the same process.' He describes a dysfunction of this system with an equally Irish meta-phor: the 'Unholy Trinity' of dysregulated stress, inflammation and dysbiosis.

You are the host of between 20 and 40 trillion microorganisms. This is made up of tens of thousands of species of bacteria, viruses, fungi and other types of microscopic organisms. Your body is their world. But this is just a drop in the ocean of nature: there may well be a trillion species of microbes on Earth, their number totalling a nonil-lion (1,000,000,000,000,000,000,000,000,000,000). There might be more microbes on Earth than there are stars in the universe. When I think about this profusion of life, I feel a profound sense of wonder. And this wonder might do us wonders: a young body of research is starting to explore how wonder and awe can be beneficial for our minds and bodies.[48] This has been explored through psych-edelics, spiritual experiences and spending time in nature, but this should also apply to the beautiful complexity of your own body. And you are the steward of an ecosystem. You are part of something much larger than yourself. If you look after your microbiome, you are tending a delightful, blossoming garden. And if you look after your ecosystem, it will look after you.

There are many practical ways in which you can nurture your gut microbiome for your body and mind. Later, we will examine the exciting evidence behind how diet (and in some cases, targeted psychobiotics) can improve our mental health. We will see how Metchnikoff really was on to something when he took an interest in Bulgarian peasants and their penchant for fermented yoghurt. But for the time being, let's simply take stock of the recent revolution in our understanding of mind and body. In the last few chapters we have seen that, far from the brain being encased far away from the messiness of the immune system, our brain and our mind are utterly intertwined with our immune army and the microscopic civilizations we harbour. Now that we are able to conceptualize these as one brilliant, cohesive system – our defence system – we may begin to understand what happens when this system goes wrong.

PART TWO

Things Fall Apart

6

Friendly Fire

When the immune system turns on its head

Et tu, Brute?

WILLIAM SHAKESPEARE,
JULIUS CAESAR

The Mexican cartel and the immaculate conception

Samantha Raggio had every reason to be happy. It was late 2019 and the twenty-nine-year-old Californian had moved back to her home town with her husband. Angels Camp is a picturesque Gold Rush town nestled in the foothills of the Sierra Nevada. 'It's a small place,' she told me. 'Perhaps a hundred people graduated from high school with me ... but I like it, and it's where my family has always been.' I scrolled through images of the town online as we chatted via video-link. It fits every stereotype a Brit like me has of the American frontier: a main street walled by a mishmash of eighteenth-century facades, a riot of wooden balconies and colourful awnings adorning banks, bars and emporiums of various kinds. I was half-expecting one of the images to show a brooding outlaw leaning on a wooden pillar outside the saloon, waiting for the sheriff's return. Samantha needed this homecoming: she now lived close to her precious nieces, had thrown off the weight of a stressful marketing job to become a nanny, and she was finally able to spend more time on her real passion: designing and making jewellery.

But something was wrong. 'For months I had been feeling rough. I was depressed, had this really bad fatigue and it felt like I constantly had the flu. I went to the doctor's a number of times but it didn't really come to anything.' One morning in October 2019, things started to get really strange. Embarking on making a new earring, she started to roll out some clay. After her right hand had squeezed the material into a ball, it would not relinquish its grip: 'I'd completely lost control of my hand.' Later on that day, when she was midway through lifting her fork to her mouth during lunch, that same hand ignored her will and let go of the fork. It was clear that something was amiss, but it took a couple of days for Samantha to realize that things were very wrong indeed. Her mother visited, asking for advice on where in her living room she should hang a new picture. 'Mom,' Samantha replied, 'what does your living room look like?' She had to be reminded that this was, in fact, the living room of her childhood home. Sensing that she might be losing her grip on reality, Samantha called her doctor, who advised her to go straight to the emergency department.

Not that the trip to hospital helped. 'The blood tests were normal, and they could see on my records that I have a history of anxiety and panic attacks. They must've assumed that whatever was happening was psychological. They sent me home with no plan.' Samantha came back home and lay in bed. She was physically alone – her husband was away with work – and had never felt so isolated. She was losing control of her body and had now started to lose her memory. More than that, she felt that she was losing her identity, and no one seemed able or willing to help her find it.

It had been a long day, and Samantha suddenly realized that she was very thirsty. True, she hadn't had much to drink, but there was something ominous, yet intangible, about this sensation. She got up out of bed and approached the kitchen in search of water. Now at the sink, she began to reach for a cup. Suddenly, out of the blue, both of her fists clenched shut and her arms flexed inwards. Her elbows folded automatically, like a pocketknife snapping shut. For the briefest of moments, Samantha knew that she was going to fall. Then she passed out.

'That wasn't the weirdest thing, though,' Samantha recalls. She woke up a few minutes later, spreadeagled on the kitchen floor. But, inexplicably, she was lying about eight feet away from where she fell. Her tongue felt sore and heavy, and her ankle throbbed in agony. It was as though she had been possessed. In a daze, she managed to call an ambulance and was rushed to hospital. Some unclear amount of time into her admission – Samantha is only just piecing back her memory of this period – a white-coated neurologist came over to her bed. He looked at her and briefly peered through her notes before asking her to perform various movements with her limbs and facial muscles. Whipping a pen torch out of his coat's top pocket, he inspected her swollen tongue, blackened by biting. He noted her clearly broken ankle. In the same neutral tone he used to examine her, the neurologist delivered his diagnosis: 'You've had a seizure. We'll be seeing you in a few days for some brain scans.'

The following day, Samantha was recuperating on the sofa in her mother's living room. The TV in the corner of the room hummed with the background babble of advertisements and news broadcasts. But something caught Samantha's attention. The newsreader was talking about the aftermath of a mass shooting carried out by a Mexican drug cartel, somewhere near the US border. As the scenes of carnage were broadcast, Samantha began to feel a sense of dread: 'Don't I know those bodies?' Just as if the television set was reading her mind, the camera zoomed in on the bullet-ridden victims, revealing in graphic detail what appeared to be the bodies of her sister and two-year-old niece. Was this a coincidence, or were the drug cartels after the whole of her family? With chilling certainty, it dawned on Samantha: the drug lords were out to kill every last one of her relatives. And not only this; she knew, she just *knew*, that the leader of the drug cartel was her best friend's husband. Samantha's train of thought was broken by the sound of her mother in the kitchen, grabbing dinner plates from the cupboard. 'Mom!' she hissed. 'Be quiet! They'll hear you. They're outside.'

Samantha's next memory takes her back into the neurologist's office. Her new hallucinations and delusions had rallied her family to take her to hospital once again in search of answers. The neurologist nonchalantly updated his diagnosis: 'The brain scans we've done today are completely normal. We did an MRI scan that looks at the structure of the brain and an EEG (electroencephalogram) that shows us the brain's electrical activity. What Samantha is experiencing,' the neurologist told her mother, 'is a psychotic break. I think she has schizophrenia. She needs to see a psychiatrist.'

In the space of a few days, Samantha's diagnosis had been updated from anxiety attacks to epilepsy, and now to schizophrenia. But before we venture any further, we need brief definitions for the oft-misunderstood terms of 'psychotic' and 'schizophrenia'. Psychosis is an experience in which you are unable to accurately perceive reality, which manifests in abnormal perceptions (hallucinations), fixed, false beliefs about the causes of perceptions (delusions), and disordered thinking. Numerous things can cause a psychotic episode, from extreme psychological stress to psychoactive drugs. Schizophrenia is, essentially, a condition characterized by chronic (long-term) psychosis, whether continuous or episodic. It has absolutely nothing to do with having multiple personalities, as is frequently assumed in popular culture. Once chronic, very few recover completely, and it seemed that Samantha was beginning to head down this path.

Her next relatively clear memory lands us in a psychiatrist's office the following day. She can't remember what the psychiatrist looked like – her memories become particularly hard to retrieve from the time she descended into psychosis – but she remembers the doctor's attributes: a warm, measured woman. 'This isn't schizophrenia,' she told Samantha. 'You need to be properly seen by the neurologists.'

Samantha was being bounced between neurology and psychiatry – from brain to mind and back again. And all the time, her grip on reality was slipping. Back at home, she was becoming increasingly agitated and paranoid. For yet another desperate hospital visit, Samantha was loaded into the back of a family member's van. On their way to UC Davis Medical Center in nearby Sacramento, she felt a trickle of fluid run down her inner thigh. 'My waters are breaking!' Samantha cried. She had suddenly become convinced that she was pregnant with a miracle baby. And not any old baby: in Samantha's mind it seemed a reasonable assertion that her child was special: 'perhaps even Mary, the mother of God'. In reality, she was losing control of her bladder function and had wet herself.

Her last terror-filled memories are those of her sitting in the hospital's emergency department. She could vividly hear the crack of gunshots just outside the building. It was her friend's husband – the leader of the Mexican drug cartel – lining up members of her family and shooting them, one by one.

Samantha's memories return at a point several days later. By this time she had been correctly diagnosed and had started to receive identity-recovering, life-saving treatment – treatment that had stopped her descent into madness. As she came around, Samantha began to pick up cryptic clues as to what had happened. The green markings on the whiteboard at the opposite end of her small hospital room – an indecipherable scrawl of messy doctors' handwriting – might as well have been hieroglyphs, but she could make out a cluster of four large capitalized letters: 'NMDA'. A pair of young doctors, perhaps medical students, shuffled past her door, briefly peering in. As they passed, Samantha could make out words of their conversation: '. . . brain on fire'.

A new disease

In 2005 Dr Josep Dalmau, a Catalan neurologist working at the University of Pennsylvania, had his curiosity piqued by some mysterious cases. He had identified four young women who, despite the attention of legions of medical specialists, had a seemingly undiagnosable disease.[1] They all presented with a relatively rapid onset of similar symptoms: hallucinations, delusions, memory problems and seizures. This cocktail of sudden psychosis, personality change and uncontrollable movements would once have been put down to the effects of magic, spirit possession and witchcraft. Four thousand years ago, the ancient Egyptians attributed these symptoms to a wandering uterus, a disease process embraced by the ancient Greeks and Romans. Hysteria, a word stemming from the Greek word for uterus – *hystera* – became a catch-all diagnosis for odd behavioural and neurological changes in young women right up into the 1900s. Even at the end of the multiple medical revolutions of the twentieth century, Dr Dalmau was faced with four young women whom science could not explain.

But he had a hunch. They all appeared to have encephalitis: inflammation of the brain. This is most commonly caused by a viral or bacterial infection, but in these four cases multiple investigations found no infectious cause. What was also strange was the predominance of psychiatric symptoms compared with what you would expect in viral or bacterial encephalitis. A couple of clues, however, suggested an alternative diagnosis.

First, spinal taps did indeed reveal signs of inflammation in the cerebrospinal fluid. Second, all four patients happened to have benign, congenital tumours on their ovaries, called teratomas. A teratoma is an odd thing. All of our tissues and organs derive from stem cells, remarkable little blank slates that can become almost anything in the body. Usually the differentiation of stem cells into specific cells and tissues is an ordered, if complex, process. Teratomas, meanwhile, are the result of a disorganized, uncontrolled explosion of tissue growth, the resulting mass being an ugly mélange of different tissues: it might be hairy, contain bits of muscle or brain tissue, or even sport fully formed teeth. Dr Dalmau hypothesized that all four of these women must have had some brain tissue form within their teratomas, and that the immune system of each woman's body had detected this alien growth. Antibodies were then produced against the tumour, sticking to the abnormal tissue and flagging it for destruction by immune cells. This is all well and good, but once these women's immune systems had started to produce antibodies against brain tissue found in the tumours, these new antibodies also homed in on the same targets in their own brain.

Dr Dalmau's hunch turned out to be right. It seemed that an autoimmune process had been to blame, as symptoms improved after the women were given steroids, which dampen down the immune system's response. Now he had to find the specific target of the antibodies' ire. After almost a year of trial-and-error tinkering in the lab, dropping samples of the women's spinal fluid on to frozen sections of rat brain, he found that the antibodies bound to one specific receptor: the NMDA receptor.[2] Through what Dr Dalmau describes as 'a product of serendipity and effort',[3] a new disease had been discovered: anti-NMDA receptor encephalitis.

NMDA (N-methyl-D-aspartate acid) receptors reside on the surface of synapses – the tiny gaps between nerve cells in the brain. They are found all over the brain, but in particularly high concentrations in the hippocampus (the seat of memory formation) and the frontal lobe (crucial for the development and regulation of personality, emotions and problem-solving). NMDA receptors are activated by a molecule called glutamate, which is critical for our brain's ability to change and adapt to the constant barrage of stimuli pouring in from the outside world. We've seen that the brain is a remarkable, predictive machine that is constantly refining its model of the outside world by comparing external stimuli to the brain's internal predictions. This process is dependent on properly functioning NMDA receptors. If those are knocked out, these fine adjustments cannot be made. And if we are not able to correctly update our brain's predictions about the world, we can end up paying too much attention to the sensory noise in the environment, assigning excessive importance to irrelevant stimuli. We begin to lose our ability to 'test' reality, ultimately resulting in hallucinations and delusions. Hence Samantha watching the images of a drug cartel's victims on TV and then believing that the bodies belong to her relatives, or interpreting the urine running down her leg as her waters breaking.

Other evidence for NMDA's crucial role in reality testing can be attested by anyone who has been in a 'K-hole'; the recreational drugs ketamine and PCP ('angel dust') bring about their hallucinogenic effects by blocking this receptor. People like Samantha, and the remarkable work done by scientists such as Josep Dalmau, have helped to reveal that NMDA receptors might well be the doors of perception. They've also demonstrated the bizarre fact that these doors of perception can be bolted shut by an over-zealous immune system. Anti-NMDA encephalitis is a case of the immune system attacking the mind; it is one part of the defence system stabbing another in the back. The 217th recorded case of anti-NMDA encephalitis, which occurred just a few years after the 2008 publication of Dalmau's discovery, put this new disease firmly on the map.

Brain on fire

Susannah Cahalan's story is remarkably similar to Samantha Raggio's. For most twenty-four-year-olds in the mid-recession New York of 2009, getting a rewarding job was a distant dream. But as a young journalist, Susannah had a nose for news. She soon landed a job as a reporter for the *New York Post*, bought an apartment in Manhattan's West Side and found herself partying with the city's gilded youth. But one morning, Susannah woke up from an intense dream about bedbugs and discovered that her bed – in fact, her whole apartment – seemed to be crawling with them. She called in the exterminator and, despite his protestations that there was not a single sign of infestation, ordered him to fumigate the place.

This delusion of infestation was the first sign of a rapid descent into her 'month of madness', which she recounts beautifully in her 2012 memoir *Brain on Fire*.[4] She began to scour her boyfriend's emails for signs of infidelity, became convinced that her father had been replaced by an identical but evil imposter, and believed that news reporters on the TV were talking about her. She also experienced a number of seizures. For weeks, Susannah was bounced between specialists, collecting a sticker book of misdiagnoses: bipolar disorder, schizo-affective disorder, alcohol withdrawal and, to quote one neurologist, 'partying too hard, not sleeping enough and working too hard'.

Admission to hospital brought doctors no closer to identifying the cause: all blood tests and brain scans were normal. A psychiatric hospital beckoned. That was until she was seen by Dr Souhel Najjar, a New York-based Syrian-American neurologist who had been keeping abreast of Dr Dalmau's recent discoveries in nearby Phila-delphia. He suspected an autoimmune cause and requested a biopsy of her brain. This showed immune cells attacking Susannah's brain, with antibodies targeting her NMDA receptors. With a level of treachery worthy of the Roman senate, her brain was being attacked by what she called her 'Brutus body'.

Dr Najjar suggested a 'three-pronged attack'. This began with steroids, a crude tool to reduce inflammation in the short term. She then underwent plasma exchange: a dialysis-like procedure in which her rogue immune cells were washed out of her blood and replaced. Finally, she would come back to hospital on a monthly basis to have intravenous immunoglobulin (IVIG), an infusion of donor anti-bodies that 'mop up' autoreactive antibodies in the blood. Susannah made a full recovery.

The decade following *Brain on Fire* has shown that Susannah Cahalan was just the tip of the iceberg. I decided to talk to one of the world-leading experts in the field of autoimmune encephalitis: Professor Belinda Lennox. As well as being Head of the Department of Psychiatry at the University of Oxford, she is also my research supervisor and colleague.

'In the first few years after the discovery of anti-NMDA receptor encephalitis in 2008, interest in autoimmune encephalitis spread rapidly,' she tells me when we meet in her office. 'It was amazing: when previously we were faced with cases of predominantly young women who rapidly developed psychosis and often died, we had now identified a direct cause that could be easily treated. This is unusual in any area of medicine, let alone neurology or psychiatry. It was also clear evidence for the role of NMDA in psychosis and schizophrenia, something that was gaining traction in the scientific community.'

At the time, Belinda was a Cambridge-based psychiatrist, treating individuals with recent-onset psychosis. She wondered whether some of her patients who seemed to be showing the first signs of schizophrenia were in fact suffering from this autoimmune encephalitis – particularly those who did not respond to conventional psychiatric medications. 'So, I got in touch with Angela Vincent in Oxford, who I think of as the Queen of Neuroimmunology: she was the first person to identify encephalitis caused by antibodies attacking the brain, back in 2001.'* Ten years later, with her help and advice, I collected blood samples of forty-six patients with psychosis, and it turned out that four had antibodies directed against neurons in the brain, three of which were against the NMDA receptor.'[6]

What was particularly remarkable about their study was that it included the first description of autoimmune encephalitis presenting with purely mental symptoms. While many patients – like Susannah and Samantha – exhibit neurological symptoms (such as seizures

---

* Angela's discovery of antibodies attacking potassium channels in brain neurons at the turn of the twenty-first century helped scientists like Dr Dalmau make their own groundbreaking discoveries.

and movement disorders), there is an increasing recognition that psychiatric symptoms alone may indicate the condition. In 2020, after years of intense discussion, an international consortium of neurologists and psychiatrists produced a consensus statement on defining, diagnosing and treating this disease, now termed 'auto-immune psychosis'.[7] The authors of this landmark paper included Professor Lennox and Dr Najjar, the hero from *Brain on Fire*.

Ironically, diagnosing these cases is still difficult, given the stark disparity in funding and support for those with 'mental' rather than 'physical' illness; a seizure is much more likely to get you access to expensive and intensive investigations than a hallucination, even if the same biological process is at play.

Belinda Lennox had found patients whose psychiatric symptoms were caused by the immune system attacking the mind. The disease was hiding in plain sight. When Dr Najjar correctly diagnosed Susannah Cahalan in 2009, he guessed that roughly 90 per cent of those with anti-NMDA encephalitis were undiagnosed. The general consensus today is that this percentage is probably lower, but it's still highly likely that the number of misdiagnosed cases still heavily out-weighs those successfully treated. A whole decade on from Susannah Cahalan's case, when Samantha Raggio started to believe that a Mexican drug cartel was after her family, it still took numerous reviews by medical professionals to reach the correct diagnosis. We can be certain that there are individuals with this disease today who, misdiagnosed, are missing out on life-saving treatment. As Susan-nah writes in her memoir, 'How many people currently are in psychiatric wards and nursing homes, denied the relatively simple cure of steroids, plasma exchange or more intense immunotherapy?'

In the 2010s, Belinda Lennox and her team carried out a large study to find the prevalence of brain-attacking antibodies in patients with a first episode of psychosis, with findings of around 9 per cent.[8] Subsequent studies have found similar results – between 5 and 10 per cent. In 2018, a psychiatric team from Queensland, Australia, tested the blood of 113 inpatients with psychotic disorders for anti-neuronal antibodies.[9] They found that six patients had the antibodies, five of whom were given immunotherapy – four were cured. While it's clear that antibodies are not responsible for most cases of psychosis, it is chilling to think that in every psychiatric hospital there are very likely at least a few patients with undiagnosed and treatable auto-immune psychosis. They are probably considered 'treatment resistant', not benefiting at all from conventional psychiatric medications. They have been destined for lifelong disability, a tragedy deepened by the fact that the means to complete recovery is out there. And there are also likely to be many individuals suffering from autoimmune condi-tions we do not yet know exist. NMDA is not the only target of rogue antibodies in the brain: as I write in 2023, antibodies directed against more than twenty different types of receptor or molecule in the brain have been found in individuals presenting with psychiatric symptoms. There are certainly more to be found.

If there are still plenty of cases of autoimmune psychosis hiding in plain sight today, it is sobering to think that autoimmune psych-osis has gone undiagnosed up to now. Reading some of the first descriptions of schizophrenia by the German psychiatrist Emil Kraepelin in the early twentieth century – psychiatric symptoms mixed with seizures, fluctuating confusion and odd, catatonic movements – it is clear that some of these would have had auto-immune causes.[10] Perhaps, with the benefit of hindsight, we can even solve historical mysteries.

That might include a mysterious event in Massachusetts in the early spring of 1692. Deodat Lawson, the former minister of Salem Village, put pen to paper to try to make sense of what he had just seen. Lawson had been invited to lead a service in the village, and after the singing of the first hymn, eleven-year-old Abigail Williams had suddenly shouted, 'Now stand up, and name your text!' Startled, but unwavering, Lawson began reading from scripture. When he had finished, young Abigail's voice rang out again from among the congregation: 'That was a long text!' Her behaviour was certainly indecorous, but he would never have expected it to become malevolent. He recorded the subsequent behaviour of Abigail and her nine-year-old cousin, Betty Parris, over the following days:

'She was at first hurried with violence to and fro in the room . . . and begun to throw fire brands, about the house; and run against the back, as if she would run up the chimney . . .'

'Sometimes making as if she would fly, stretching up her arms as high as she could, and crying Whish, Whish, Whish! Several times . . .'

'They had several sore fits . . .'[11]

Their symptoms – sudden and extreme behavioural changes, hallucinations and convulsions – contained all the ingredients of the supernatural. Witchcraft was considered the obvious cause, and it was not long until the scapegoats were identified. Three women, all social outcasts, were accused of using magic to afflict the girls: Tituba (an enslaved woman with no given surname), Sarah Osborne and Sarah Good. While accusations of witchcraft were dying out in Europe by the late seventeenth century, they were still not unusual in the New World. What was unprecedented, however, was the subsequent eruption of paranoia that became perhaps the most famous case of mass hysteria in history: the Salem Witch Trials. Within a year of Betty's and Abigail's symptoms, over two hundred people had been accused of witchcraft and twenty had been executed.

A growing number of scientists and historians argue that unsolved historical cases of bizarre behavioural symptoms may well be explained by autoimmune processes, whether it be a seemingly possessed young girl in seventeenth-century Salem,[12] or, as another paper suggests, the fourteen-year-old Missouri boy whose 1949 case inspired *The Exorcist*.[13] Although we don't have a time machine and will never know for sure, we can be certain of this: autoimmune psychosis has ruined or ended the lives of many, many people throughout history, and until the twenty-first century, all their cases were misdiagnosed. From exorcism to conventional psychiatric medication, cures may well have been given in good faith but without any knowledge of the disease process at all. And in an ironic – even perverse – quirk of biology, anti-NMDA receptor encephalitis may have sustained the belief surrounding 'hysterical' women as well. Throughout history it has not been an uncommon occurrence for doctors (and they were invariably all men) to remove the uterus of a woman presenting with 'hysteria'. In some cases this must have worked, but not in the way they hypothesized. By unknowingly removing an ovarian teratoma – the tumour that sparks the formation of anti-NMDA antibodies in some cases – these surgeons had unintentionally stopped the production of the rogue antibodies that were targeting both tumour and brain.

If there is a single lesson in all of this, it is that we should be very humble about our assumptions of the causes behind symptoms, even when we draw on long-established medical wisdom. The concept of autoimmune psychosis was inconceivable a decade or so ago. Who knows what else we are about to discover.

The good news is that autoimmune psychosis is being picked up more regularly due to increased awareness and higher-quality blood tests, and immune treatments appear to be effective. A German study, published in 2022, examined the records of ninety-one patients with autoimmune psychosis, and found that immunotherapy worked in 90 per cent of people with confirmed antibodies in the blood.[14] Even for those without antibodies in the blood, but where doctors suspected an autoimmune cause via other tests (such as an MRI scan), there was an 80 per cent improvement following immunotherapy. One particularly successful combination is intravenous immunoglobulin (IVIG) and rituximab, a drug that deactivates antibody-producing cells.

Belinda Lennox is currently leading the ultimate test of this treatment: a randomized controlled trial. The SINAPPS2 trial involves screening thousands of patients presenting at mental health services with psychotic symptoms for anti-neuronal antibodies.[15] The patients are then randomized, and either given a placebo or the cocktail of IVIG and rituximab. It was this combination of drugs that led to Samantha Raggio's recovery, and it will hopefully save the lives of many others. The current drugs we have for treating psychosis only dampen down symptoms; these immune treatments – for those with autoimmune psychosis – can bring about a cure.

Of causes

Autoimmune psychosis is perhaps the most extreme example of the immune system influencing the mind. It is also perhaps the most clear-cut in terms of cause: an aberrant antibody attacking brain tissue. In the next chapters we will discover that many other neuro-psychiatric disorders are caused or worsened by an immune system gone rogue, from depression to dementia. We will see that, in most cases, we find evidence of a supersystem out of balance. Our defence system – both mind and body – can be thrown out of kilter by chronic inflammation, infection, psychological stress and trauma, all mixed up in a mélange of environmental and genetic influences. While this story is not as simple as Samantha's, it is one that affects far more people. In fact, in our modern world many of us probably struggle with an unbalanced defence system. But before we get to that, let's explore the fascinating implications for long-term psychotic disorders such as schizophrenia.

There is a remarkable link between psychotic disorders and auto-immune conditions. This not surprisingly includes autoimmune diseases that directly attack the brain, such as multiple sclerosis (MS) and lupus. One 2023 piece in the *Washington Post* – which became a viral sensation across social media – tells the remarkable story of April Burrell.[16] In 1995, the bright, outgoing twenty-one-year-old trainee accountant from Baltimore suddenly began to develop visual and auditory hallucinations. She was diagnosed with schizophrenia and, as her condition deteriorated, was admitted to a psychiatric hospital in 2000. It was around this point that she began to descend into catatonia: spending hours sitting or standing, muttering to herself in a trance-like state, completely oblivious to everyone around her – whether they be staff or a family member. April remained like this for twenty years until the late 2010s, when doctors – armed with a new knowledge of immune-brain interactions – tested her for autoimmune conditions. It turned out that she had neuropsychiatric lupus, a condition in which the immune system attacks the brain with antibodies. April was started on a cocktail of anti-inflammatory medications similar to those given to Samantha Raggio and Susannah Cahalan, and they started working immediately. Following a remarkable recovery, she was discharged from the psychiatric hospital in 2020. After two lost decades, April regained her memory, her personality, and her life.

But it is also clear that psychosis is associated with many autoimmune conditions that are not considered brain diseases at all: psoriasis, coeliac disease and autoimmune thyroid disease.[17] A large study of almost 40,000 Danes found that those diagnosed with schizophrenia had a 45 per cent increased likelihood of having previously been hospitalized with an autoimmune disease, compared with healthy controls.[18] A similar Taiwanese study found a 72 per cent increase.[19] On the flipside, another large Danish study found that those with schizophrenia have a 53 per cent increased chance of developing an autoimmune disease.[20] What could explain this link, if it isn't caused by antibodies directly meddling with the brain?

One clue is in the realm of infections. For many decades, epidemiologists have observed a rather unusual risk factor for schizophrenia: you are more likely to develop the condition if you are born in winter or early spring.[21] One hypothesis is that infections contracted in colder months – whether in the mother, foetus or newborn baby – cause changes in the early neurodevelopment of the child. Science has caught up with this theory and largely supports it: there is compelling evidence that maternal infection is associated with the onset of schizophrenia later on in life.[22]

The fact that various different infections and autoimmune diseases increase the risk of schizophrenia strongly implicates a common mechanism as the culprit: inflammation. In earlier chapters we saw that inflammation is a crucial arm of our body's defence system: it is the immune system's coordinated response to infection and injury. It is a savage symphony of complex biological processes. Inflammation can be life-saving when it is short-lived and well regulated. But when inflammation becomes chronic it can wreak havoc. Overall, it looks like long-term inflammation in body and brain can contribute to the development of psychotic disorders.[23] Many different lines of evidence support this. First, there is a clear association between these conditions and raised levels of pro-inflammatory cytokines circulating in the blood.[24] Perhaps one link between psychotic and autoimmune diseases is the fact that infections and chronic inflammation contribute to the breaking down of the blood-brain barrier – the wall that separates brain tissue from the body's blood vessels – which increases the ability for antibodies to reach the brain.[25] Alternatively, some infectious agents that directly attack the brain can trigger autoimmune psychosis. Having a previous episode of encephalitis caused by the herpes simplex virus greatly increases one's risk of developing anti-NMDA receptor encephalitis.[26]

Alongside the mind being assaulted by infections and an overactive immune system, we must never forget the role of another part of our defence system: the gut microbiome. While research is still in its infancy, there is some evidence that a disrupted gut microbiome (dysbiosis) is associated with psychotic disorders such as schizophrenia.[27] It is well established that gut dysbiosis in early life can lead to a hypersensitive immune system, and that dysbiosis throughout life can lead to inflammation. It might also be the case that a microbiome with little bacterial diversity may not be able to produce the molecules required to form a healthy brain, particularly during the first months and years of brain development.

We've focused in on environmental causes of schizophrenia, but we should bear in mind that genetics also play a role. Schizophrenia has a heritability of around 80 per cent, meaning that genetic forces tend to have a more powerful (but certainly not total) influence on the development of schizophrenia compared with environmental factors. Interestingly, in 2009 the International Schizophrenia Consortium published the findings of one of the largest genetic studies on the disease to date, and found that the genes most strongly associated with schizophrenia were deeply involved with the immune system.[28] We are yet to tease out the exact mechanisms between these genetic variants and disease, but current theories propose that schizophrenia is more likely to develop in those who are genetically susceptible to infections,[29] as well as those who are more genetically likely to have increased inflammation.[30]

A picture is starting to emerge: aside from the rare examples of psychosis being a clear-cut autoimmune disease, it appears that both psychotic disorders and autoimmune diseases stem from a combination of infections, an unbalanced gut microbiome and genetic susceptibility, ultimately leading to chronic inflammation. One does not need to have all of these risk factors to develop disease, but there seems to be a cumulative risk. You could call it a 'multi-hit hypothesis': perhaps in one individual neither genetic susceptibility nor receiving in-utero infections alone would lead to schizophrenia, but the combination of the two tips them over the edge.

There is, however, one 'hit' that we have missed out. All of these examples are 'bottom-up': the body influencing the brain and mind. But the mind also wields a powerful influence on the brain and body: psychology is biology. It is well known that psychological stress, whether in early life or in adulthood, is associated with a subsequent development of psychotic disorders.[31] Stress can come in many different forms, from life events such as bereavement or divorce, or from the residue of trauma from emotional, sexual or physical abuse. It is intriguing that psychological stress in and of itself can cause inflammation, negatively affect our gut microbiome, and carry with it an increased risk of autoimmune disorders.[32]

This leads us to a key principle in understanding how the immune system influences mental health: there is usually no single cause. Samantha and Susannah are exceptions, but in most cases it is a combination of factors that produce disease, once they have reached a critical mass. It is a mix of genes and environment, of immune and non-immune, of top-down and bottom-up. Some clinicians and researchers picture this through the 'stress-vulnerability model': we see psychiatric disease when stress (both physical and psychological) is added to vulnerabilities (such as genetic predisposition). A commonly used analogy is that of a cup. Each of us is represented by an empty cup of the same volume. Each of us is given a different volume of liquid by our parents: our genetic vulnerability. That is rarely enough on its own to cause psychiatric or autoimmune disease. But as you pour in infections, gut dysbiosis, psychological stress or trauma, the level rises. The liquid is now a cocktail of inflammation, and disease is at the point at which the cup begins to overflow.

While we still talk about the 'mind' and the 'body', as well as 'mental' and 'physical' health, inflammation does not differentiate. Chronic inflammation results from both the mental and the physical, and it affects both physical and mental health, often at the same time. In short, it is the result of an overactive, overprotective supersystem; chronic inflammation is the wasteful, destructive heat generated by the friction of an unbalanced defence system.

We humans are meaning-making machines, and we love stories with one identifiable villain, and one identifiable hero that can vanquish them – this is what makes Samantha's and Susannah's stories so compelling. The truth about most disorders of mind and body, however, is that their causality is multifaceted and circular; vicious cycles in which mind and body, cause and effect, become indistinguishable.

This is why I believe that we should always have in mind the defence system – that mind-immune-microbiome structure we explored in Part I – when we approach chronic diseases. Sometimes the cause of an unbalanced defence system comes primarily from the body, sometimes primarily the mind, but often it is a combination of both. This new framework will help us answer some of the huge questions facing the field of mental health, which we will explore in the following chapters. Could people living with depression have chronic, low-grade inflammation that, if treated, could drastically improve their mental health? How can psychosocial stress change your immune system and damage your body? How many people have dementia that could have been prevented with anti-inflammatory lifestyle choices or immune-directed treatments?

Future generations will most likely look back on our current knowledge of the brain and our treatment of mental illness with the same pity we have for the judges and clergy of seventeenth-century Salem. That is not to discourage us: we are now beginning to shed light on the scientific darkness of the brain-immune relationship. What excites me is the discovery of how mind and body interact on an intimate level, leading to new treatments for physical and mental health conditions. Conventional psychiatric treatments work for some people, but for most they are far from a cure. The new science of the immune-mind connection is not just revealing new drug treatments – such as the antibody-fighting medications that cured Samantha's psychosis – but it is showing how relatively simple lifestyle changes can drastically improve our health.

Samantha's case demonstrates the power of the immune system to bring about extreme psychiatric symptoms. While every new case discovered and every life saved is a triumph of modern medicine, autoimmune psychosis is uncommon. A much more familiar mental health disorder – in fact, one of the leading causes of disability worldwide – is depression. It's time to explore how the immune system is an unexpected culprit in this common condition.

7

The Inflamed Mind

Inflammation and depression

To feel depressed, I went to Cardiff . . . uh, what's the point . . .

A welcome side effect

If I asked you to name a body part, outside of the skull, that holds the key to the treatment of depression, what would you choose? Having read the book so far, perhaps you might have chosen the gut, or even an immune organ such as lymph nodes, bone marrow or the spleen. I'm sure that you could make great arguments for other bits of the body. But I'd bet my house on the fact that you did not choose a finger joint.

I came to this left-field answer after speaking with Professor Iain McInnes, a Glasgow-based world-renowned rheumatologist. Rheumatology is curious: it has almost no hold in the collective consciousness, yet it is an immensely important specialty, practised by some of the sharpest minds in medicine. It is commonly thought of as a specialty of joints and bones, but it is much more than that: rheumatologists specialize in diagnosing chronic autoimmune and inflammatory diseases that affect every tissue in the body, from arthritis to vasculitis. Rheumatologists are also responsible for dealing with shape-shifting, whole-body conditions of outrageous complexity, such as lupus and rare connective tissue diseases.

There is much to talk about with McInnes, from an academic and clinical career that has brought with it the presidency of Europe's leading rheumatology society, leadership of Glasgow's medical school and a CBE – but I was most interested in his early days as a rheumatologist. At the turn of the millennium, McInnes was working in a rheumatoid arthritis clinic at the Glasgow Royal Infirmary, a venerable institution in the city's East End – what was then one of the most deprived communities in Western Europe. Rheumatoid arthritis is a chronic condition in which the immune system starts attacking the joints, most commonly in the hands and wrists. McInnes would see patient after patient, many in early middle age, with fingers so swollen and inflamed that they could barely pick anything up. Chronic joint erosion leads to grossly deformed hands, rendered useless and painful. Many of McInnes's patients had not responded to any of the treatments offered to them, and they were understandably miserable. But rheumatology was on the cusp of a golden age.

A brand-new class of drugs, called biologics, had recently entered the clinical world. A biologic is essentially a lab-made protein, such as an antibody, that has been designed to target a very specific molecule in the body. Biologics represented a new generation of anti-inflammatories: if aspirin and steroids are shotguns, biologics are laser-guided rifles. McInnes was able to offer two of these biologics to his treatment-resistant patients, each of which targets and neutralizes an inflammatory cytokine (one zoned in on tumour necrosis factor alpha, or TNF-α, and the other on interleukin 6, or IL-6).

'In the initial trials, we were blinded as to whether we were giving patients the placebo or the biologic,' McInnes tells me, 'but it soon became clear who was receiving what – the biologics worked incredibly well.' For many patients, previously viewed as untreatable, these biologics really were a silver bullet. 'We were delighted … and, of course, the patients were delighted too.' You certainly would be delighted if your swollen, bulging joints had suddenly started to heal.

As with any new drug, McInnes was on the lookout for side effects. One common one that he noted was remarkable: 'Some of my patients reported a significant improvement in their mood, describing themselves as feeling as good as they had done for years. But when I examined their hands, their fingers were just as swollen as before; their knuckles just as deformed.' In some patients, these new anti-inflammatory drugs seemed to lift mood and relieve depression, even if they did not deliver their intended outcome for 'physical' symptoms.

McInnes's other interest at the time – which continues to this day – is psoriatic arthritis: chronic joint-swelling accompanied by the red, scaly plaques of psoriasis. Again, biologics were often revolutionary for these patients, soothing joints and clearing up skin plaques like bleach removing stains. 'For some people, however, it just didn't work. But, regardless, these people would still report feeling better in themselves, even when their visible, disfiguring skin disease hadn't receded one bit!' Brain fog lifted, fatigue reduced and mood improved. It was clear that psychological improvements were not simply a result of improvement in joint and skin symptoms. 'And it's important to remember that these were people who came from deprived backgrounds in the impoverished East End who had dealt with real adversity in their lives. They weren't just being polite about the new drugs.' These patients were not afraid to speak their minds, and they were clear that these new immunotherapies were dramatically improving their mood. Iain's patients were offering a glimpse into the future of mental health. Perhaps, for some people at least, depression and arthritis both resulted from the same root cause: inflammation.

Around the same time that joint specialists like McInnes were unintentionally curing depression in some of their patients, liver specialists found that they were doing the exact opposite. Hepatitis C is a pernicious disease. When this tiny, spherical virus infects a human host – often via blood-to-blood contact – it usually brings about only mild infectious symptoms (or even none at all) as it sets up home in the liver. The hepatitis C virus is, however, a terrible tenant, and over the course of years it causes cirrhosis and brings a greatly increased risk of deadly liver cancers. In the early 1990s, doctors were starting to use a treatment that proved very effective: interferon-alpha (IFN-α). This is a protein naturally produced by cells of the immune system to initiate an antiviral immune response, and researchers found that giving it in artificially high doses helped stimulate the immune system to evict the liver's unwelcome lodger. There was a catch, though. Almost half of the patients given this pro-inflammatory potion became depressed.[1] Many continued to experience sustained and recurrent depressive episodes long after the treatment had finished.

These unintended side effects raise profound questions. Can inflammation cause depression? Could treating inflammation cure depression? Before we even think about starting to address these questions, we need to define our terms. Almost every concept and condition in the realm of mental health is contested and caricatured, and depression is no different. We all experience dips in our mood; indeed, this is normal and adaptive – and that is not what I mean by depression. Also known as major depressive disorder (MDD), clinical depression is a severe mood disorder in which someone is low in mood or has no pleasure or interest in doing things for most of the day, most days, for at least two weeks. Depressive episodes – particularly if untreated – usually last much longer. It isn't simply feeling sad; it's a chronic, whole-body experience. Patients often describe it in physical terms: having a ball and chain tied around your leg; lying trapped beneath a boulder; treading water in the middle of the ocean. The two core symptoms of depression are low mood and anhedonia. This word, unfamiliar to most, conveys such an important experience: an inability to feel pleasure from things that you usually enjoy: hobbies, food, sex, and so on. The world becomes a futureless, hopeless shade of grey. Alongside these core symptoms, other common symptoms take hold on the body: fatigue, slowness of movement and mind, insomnia or hypersomnia (sleeping too much), weight loss or gain, feeling guilty or worthless, impaired concentration, and thoughts of suicide.

Depression is terrible, and it is common. Roughly one in seven people will experience at least one full-blown depressive episode in their lifetime.[2] It is a tragedy on an individual and a societal level, yet we – and by we, I mean all of us, including doctors – struggle to grasp it. This is because depression is not one disease; it is a constellation of symptoms that appears similar across sufferers, despite resulting from a melting pot of diverse causes and mechanisms. There is no single reason for depression: one person's depressive episode seems to clearly stem from recent life events, another from historic trauma and loss, yet another out of the blue, with no clear psychological or social cause whatsoever. There is also no one presentation of depression: in fact, there are 227 possible symptom combinations that meet the diagnostic criteria for major depressive disorder.[3] Perhaps understandably, there is no one treatment for this condition, either. Antidepressants lead to remission in roughly a third of patients, offer some kind of improvement in another third, while a final third remain completely unaffected.[4] There are similar findings with psychological interventions. This is not to mention the withdrawal symptoms and side effects associated with antidepressant medications. I have seen the near miraculous effects of antidepressants on some individuals, but there is little way of knowing in advance who will benefit, and many well-meaning clinicians end up giving individuals a treatment that makes their life significantly worse.

So, we say that depression is 'heterogeneous': diverse in cause, presentation and treatment response. The goal of many clinicians and researchers, therefore, is 'stratification': dividing depression into different subgroups. One subdivision that has gained particular interest over the last decade or so is 'inflamed depression'. Perhaps the best-known proponent of the idea that inflammation can cause depression is the Cambridge neuropsychiatrist Professor Ed Bullmore, who outlined this thesis in his 2018 book *The Inflamed Mind*. In the relatively short time since the publication of his book, much more evidence of the existence of inflamed depression has emerged. Around a quarter of people living with depression have mildly raised levels of a commonly measured inflammatory marker called C-reactive protein (CRP) in their blood, suggestive of low-level, chronic inflammation.[5] Other inflammatory molecules are also raised in the blood of individuals with clinical depression,[6] and a 2022 study found robust evidence for increased numbers of numerous types of immune cells in the blood of sufferers.[7] This group of people – who are both depressed and have raised inflammatory markers – is a particularly curious one. They tend to respond particularly poorly to conventional antidepressant medications.[8] It also appears that inflammation is more strongly associated with some depressive symptoms than others: alongside the core symptoms of persistent low mood and anhedonia, individuals with raised inflammatory markers in the blood tend to experience greater fatigue, an increased need for sleep and more changes in appetite compared to the uninflamed. Clinicians have long known about this group of symptoms, which has historically been termed 'atypical depression', but it is only recently that we have seen clear evidence that this condition may be caused by inflammation. Could roughly a quarter of people diagnosed with depression – which would account for around 70 million people worldwide – be living with a whole-body inflammatory disorder?

It is a tantalizing thought. But let's not get ahead of ourselves. We always need to be careful not to assume that correlation means causation – between 2000 and 2009 there was a 94 per cent correlation between cheese consumption per capita and the number of people who died by becoming tangled in their bed sheets.[9] I doubt that Camembert directly causes linen entanglement. In a similar way, could the association between depression and inflammation be a sideshow? Perhaps people with chronic illnesses (which often come with inflammation) are more likely to become depressed? Or could it be that depression causes behaviours that result in inflammation – from smoking to reduced exercise? Does inflammation really cause depression, or is it simply a correlation?

Stuck on blue

To find out whether inflamed depression is a real disease entity – a cause and not an effect – we need to answer some questions. The first is this: is there a mechanism that would explain *how* inflammation causes a persistent and significant change in mood?

The most long-standing evidence for causality is sickness behaviour: an infection preceding – and clearly causing – depressive symptoms. As we discovered in Chapter 3, sickness behaviour is a normal, adaptive defence mechanism, fine-tuned across millennia to protect us from ill-intentioned microbes. An infectious agent – whether it is a seasonal influenza virus or an undercooked kebab – triggers innate immune cells: the immune system's first line of defence. As well as raising the alarm throughout the immune system, inflammatory cytokines and cells communicate with the brain to bring about energy-saving and antisocial behaviours: you become low, withdrawn and lethargic.

Let's have a quick reminder of the parallel pathways that bring this about. First, nerves located throughout the body can rapidly communicate the immune state they pick up to the brain. Second, inflammatory cytokines travel through the blood and influence the brain either by interacting with the blood-brain barrier, or by entering the brain itself. Third, there is evidence that in some cases immune cells can travel to the brain to influence behaviour.[10] Within the brain itself, microglia (your brain's resident immune cells) can become activated by any of these pathways, intensifying the inflammatory atmosphere by spewing out cytokines of their own. Sickness behaviour is a mobilization of your body's whole defence system, both neural and immune, both brain and body. It is perhaps no coincidence that the symptoms of depression specifically associated with chronically-raised inflammatory markers – fatigue, lethargy, changes in appetite, psychomotor retardation (the slowing down of physical and mental activity), an increased sensitivity to pain and sleep disturbance – are the same symptoms experienced in sickness behaviour. But sickness behaviour is usually short-lived, and a bout of the flu does not clinical depression make. This leads us to a confounding question: how, for some people, does sickness behaviour become stuck?

The first place to look is in the biology of the brain. While there are obvious limitations with the use of animal models in studying mental health – mice are notoriously difficult to interview – their behaviour under controlled conditions can still provide insights into the mechanisms behind inflamed depression. In 2014, neuroscientist Professor Scott Russo's team at Icahn School of Medicine at Mount Sinai, New York, carried out what could be termed a 'depression transplant' in mice.[11] First, they observed the levels of inflammatory cytokines in mice that were exposed to social stressors. Mice that had higher levels of the inflammatory cytokine IL-6 were less resilient to stress – exhibiting depression-like social withdrawal compared with resilient mice. When the peripheral immune cells of these inflamed, depressed mice were transplanted into other mice, these mice became much more vulnerable to depression in the face of social stress, too. So, it seems that the state of the immune system precedes psychological symptoms.

Other animal studies also show how peripheral inflammation could bring about depression. Tryptophan is an amino acid (a building block of proteins) that normally breaks down into one of the brain's vital neurotransmitters, serotonin. While there is consider-able debate as to the exact role of serotonin in depression, there is little doubt that brain pathways that use serotonin are involved in mood and behaviour.[12] Inflammatory cytokines, however, have been clearly shown to activate an enzyme (called indoleamine-2-3-dioxygenase, or IDO) that shunts tryptophan away from its serotonin-producing pathway towards the kynurenine pathway. The important detail to know about this biochemical pathway is that one of its end products is quinolinic acid, which exerts considerable depressive and even toxic effects on brain tissue.[13] These findings have also been replicated in a number of human studies.[14]

Another critical mood molecule that is affected by inflammation is dopamine. This is commonly misconstrued as a molecule of pleas-ure and enjoyment, but that's not the case. Dopamine is not for liking; it is for *wanting*. If serotonin is about satisfaction, dopamine is about desire. Dopamine makes us explore our external environ-ment, whether that be seeking a reward or simply initiating movement. Some of the ingredients of the inflammatory milieu – namely, cytokines and chemicals called reactive oxygen species (ROS) – can reduce the synthesis and release of dopamine in the brain.[15] Perhaps a reduction of this motivating molecule lies behind many of the overlapping symptoms of sickness behaviour and depression, causing the sufferer to turn inwards and avoid engaging with the world. For a couple of days, this is a good thing, sending us back to bed. For months or years, though, this can be disastrous.

The explosion of research into the mechanisms behind inflamed depression in the last decade or so has revealed that brain cells may play a significant role, too. There are very few areas of the adult brain in which new neurons are born – a process called neurogenesis – and this mainly takes place in a part of the brain called the hippocampus, this Greek name reflecting its uncanny resemblance to a seahorse. The hippocampus has long been known for its role in consolidating short-term memory into long-term, and it is this region of the brain that first succumbs to the ravages of Alzheimer's disease. What is less well known is the fact that neurogenesis in the hippocampus plays a key role in mood and emotion: if neurogenesis stops or even slows down, depression begins. It is also evident that inflammation slows down neurogenesis.[16] What is particularly fascinating is the nature of the other significant cause of impaired neurogenesis: psychological stress. We will dig deep into stress in the next chapter, but here we have a window into how 'biological' and 'psychological' triggers affect exactly the same process in the brain. Persistent psychological stress and trauma is just as 'real' as a viral infection when it comes to the effect on our brains.

But cells and cytokines can only tell us so much. To fully understand how sickness behaviour translates into depression, we need to zoom out and look at brain circuits, and ultimately to address the top stratum, of mood and behaviour. To make sense of it all, I knew just the place to go. To feel depressed, I went to Cardiff. Although I was born in England, my mother hails from South Wales, which inescapably comes with a religious support of the Welsh national rugby team. I generally associate the Welsh capital with hope – albeit hope that is often dashed whenever the result doesn't go our way. This time, however, I knew my trip to the city would contain lows of a different kind. I had elected to participate in a clinical trial looking at the role that sickness behaviour plays in depression. I knew that at best I would be receiving a placebo – sitting on a plastic hospital chair in a windowless, linoleum-lined clinic room for the best part of a day and experiencing no changes to write home about. Either that, or I'd be injected with the 'study drug'. Perhaps 'drug' is a misleading name, as this one produces the opposite-intended effect of all medicines and street drugs: it makes you feel miserable. The team in Cardiff, led by Professor Neil Harrison, are experts at injecting healthy subjects with pure lipopolysaccharide (LPS). Also known as endotoxin, we met this substance in Chapter 3: a large molecule found in the outer membrane of many types of bacteria. Innate immune cells quickly detect this molecular barcode, activating your immune system and (within an hour or so) recruiting the brain to produce sickness behaviour. Neil Harrison's team were using this carefully controlled, experimentally induced illness to study changes in brain and behaviour, and how this could lead to depression.

Harrison is a pioneer in the field of immunopsychiatry. He says that even as a teenager – long before the immune-mind link had become accepted and established – he always enjoyed straddling the awkward space between mind and body. As a junior doctor he hopped between general medicine and neurology, before switching to psychiatry and joining the Maudsley Hospital in London. After completing a PhD in Neuroscience, he spent ten years in Brighton, setting up the UK's first clinical service treating individuals with depression and bipolar disorder with evidence of raised inflammation. Harrison is now the president of the world's leading organization for the study of mind-immune interactions, the PsychoNeuro-Immunology Research Society, or PNIRS. (When I first spoke with him, embarrassment prevented me from asking how to pronounce this acronym.)

Having volunteered to play guinea pig for Harrison's clinical trial, after my first visit I was sure – and it was later confirmed – that I had received the placebo. My second trip to the lab, however, was very different. The experiments were held in the Cardiff University Brain Research Imaging Centre (or CUBRIC), a shiny behemoth of a building. A long cuboid of glass and wood, rising above a warren of terraced houses in the inner city, the CUBRIC looked like a freshly settled alien spaceship that flattened some student flats upon landing. As a world-leading brain-imaging centre, it is essentially a box of incredibly powerful magnets, including MRI scanners with magnetic fields five times as strong as those used in hospitals; I was surprised the building wasn't clad with local cutlery. Within an hour of arriving I was sitting in a clinic chair in maroon scrubs, watching a doctor inject a cocktail of purified bacterial toxins into a cannula connected to my arm. It was too late to question why I had volunteered to go through with it.

Nothing happened for about an hour. Just like the placebo day, I lay back in a sky-blue mechanical recliner chair, strapped up to a thicket of wires and monitors, and chatted away with my charming companions – Natalie, the research assistant running the study, and James, the doctor responsible for my health. But, then, very suddenly, I noticed that I had run out of things to say. Not only that, but I didn't really want to talk any more. I reached into my bag for a book, only to realize that I really couldn't muster the concentration to read. A vague sense of malaise settled over me; a coldness that was not overtly feverish – indeed, I never once spiked a high temperature, which I was later told was a more unusual response to lipopolysaccharide. My attention turned inwards, focusing on what I thought was going wrong with my body and health. It was a mindfulness body scan from hell: *Now, notice your aching calf muscles from that run you did last week, and feel how uncomfortable this chair is for your lower back* ... I later reflected on the attention we pay to our physicality, on how there is a constant stream of white noise emanating from our body all the time, but we only focus on it when needed. For example, you probably didn't notice the feeling of your legs against the chair you're sitting in, until my mentioning it right now. When my immune system was activated that morning, I paid far more attention to the state of my body – particularly negative, uncomfortable states – than I would usually do. For the rest of the morning and afternoon, my overwhelming feeling was one of slowness: of wit and of body. But I wasn't left for too long feeling sorry for myself – I had some jobs to do.

Around midday I was given an infuriating computer task to complete. A pair of differing, squiggly symbols would flash up on the screen and I would have to select one of them. After this first random decision the computer program would then let me know whether I had won a pound coin, lost a pound or broken even. This repeated for what felt like hundreds of rounds; some pairs of symbols became familiar, others not. Some patterns started to emerge and I began to win more money than I was losing, but only just. I was then transferred to an MRI scanner – one roughly twice as powerful as those found in a hospital – for a couple of hours as other researchers looked at how my brain responded to my inflammation.

Mercifully, my morose state is relatively short-lived, and when I talk to Harrison in the early evening, my mood has rebounded. I ask him what the MRI scan was for. 'We're trialling out a specific type of MRI technique called diffusion-weighted imaging to look at how the microglia within your brain respond to inflammation in the body. The methods we currently use are unreliable and involve injecting people with radiotracers, which isn't ideal. So this could be an exciting way of seeing how inflammation could affect microglia in a completely non-invasive way.' But when I ask him about the potential role of microglia and neuroinflammation in sickness behaviour and depression, his response is restrained. 'Do we need activated microglia spewing out inflammation to have any of these symptoms? Do we need inflammation in the brain? I don't think we do. Biology always has parallel pathways that bring about a similar effect. These pathways often go hand in hand, but I'm not convinced that they're all engaged every time.' This is an adaptable attribute called redundancy: as the nervous and immune systems place a premium on adaptability, the body has numerous different ways of bringing about the same outcome. Indeed, research shows that while microglia can play an important role in the brain's immune response, they are not necessary for the generation of sickness behaviour.[17] During sickness, and perhaps during inflamed depression, sometimes all three pathways – nerves, immune cells and cytokines – are in play, sometimes only two of them. Sometimes only one predominates.

Much of Harrison's work has looked at how the nerves that snake around all of the body's organs are capable of communicating powerful immunological messages to the brain: 'In some of my early studies in the noughties we found that when people are inflamed, there is an increase in activity in the insula, an area of the brain that is a representation of our internal world.'[18] This seemed to reflect my symptoms just a few hours earlier: the spotlight of my attention was directed away from the outside world and towards my body. He continues: 'Around the same time we also found that inflammation caused reduced activity in the dopamine circuits involved with movement.'[19] Again, I think back to my immune-induced slowness of movement in the clinic room. I then ask him what the frustratingly difficult computer task was about. 'Ah, now that's even more interesting. The task assesses how you learn about stimuli in the environment that are associated with either reward or punishment. What we've found is that when people are inflamed, their sensitivity to reward goes down – people are less likely to strive to seek rewarding things.[20] But what is curious is that sensitivity to punishment goes up. And our brain scans supported this, showing changes in activation in the brain's reward circuits. Overall, you become more attuned and sensitive to negative things. This includes pain.' This is certainly true of me that day – I have been without doubt more sensitive to physical pain and mental anguish.

Finally, I ask Harrison the question that hasn't left my mind: how do some people seem to get 'stuck' on sickness behaviour? Could inflammation lead to long-term depression? 'That's the big one. The short answer is that we don't fully know, and processes are likely to be different for different people. It's difficult to test experimentally, as inflaming people for long periods of time to see who develops depression is not exactly ethical. But there was an interesting group we could study. We looked at patients with hepatitis C, who were soon to be given the drug interferon, so they would have chronic immune activation anyway.[21] We know that roughly a third of people who receive this pro-inflammatory treatment end up developing major depressive disorder. So we scanned these people before their first treatment, on the day of their first treatment, and then followed them up for three months. What we found was very interesting: for some people, interferon triggered increased activity in the amygdala.' This is one of the brain's key threat detectors, which we came across in Part 1. If you notice a spider crawling into your bed sheets or spot an abusive boss approaching your desk, it is the amygdala that initiates a stress response. It is well established that people with depression often have an overactive amygdala, particularly in response to potentially negative stimuli. 'What was particularly interesting,' adds Harrison, 'was the fact that interferon, this inflammatory cytokine, activated the amygdala before people developed depression. The level of amygdala activation also predicted who would get depression later on. At a similar time, we carried out another study on people with rheumatoid arthritis who were starting an anti-inflammatory drug. This produced the opposite effect: the drug tended to dampen down amygdala reactivity.'

With all these findings, we are starting to build a more coherent picture of what's going on. When you get infected or inflamed, your brain is activated as much as your immune system is, and rapid neurological changes occur. Your movement of mind and body slows down. You become withdrawn and insular, and focus on signals coming from within your body. Your attention becomes biased to the negative and stops you either enjoying or seeking out the positive. Finally, your threat-detection system goes into overdrive, reading minor threats as mortal ones. Harrison summarizes it well: 'Inflammation nudges us towards being more sensitive to negative information. For some, this can push them into a downward spiral.'

Why do some people fall into this spiral and not others? Perhaps chronic inflammation lingers in the body and brain; perhaps the amygdala has already been primed, and this latest burst of inflammation is the straw that breaks the camel's back. This priming can be caused by any combination of threats to our defence system: infections, psychological stress, genetic predisposition, an unbalanced gut microbiome or past trauma. While the reason for falling down this spiral is always an indistinguishable mélange of what we call 'physical' and 'psychological', it is crucial to note that it isn't someone's fault for falling.

Correlation to causation

While my depressogenic experience in South Wales showed that there are many outstanding questions in the field of inflamed depression, we have found biological mechanisms that explain how inflammation could cause depression. The question is now whether these findings translate to the real world.

At the beginning of the twenty-first century, physicians such as Iain McInnes were finding out that new biologic drugs designed for inflammatory conditions like rheumatoid arthritis could treat symptoms of depression. A number of researchers have since trawled back through the results of clinical trials that tested anti-inflammatory drugs for various chronic inflammatory conditions such as rheumatoid arthritis, lupus, psoriasis, asthma and inflammatory bowel disease. They were particularly interested in trials that also reported data for symptoms of depression as a secondary outcome. The overwhelming picture is that anti-inflammatory drugs – from older, generalized drugs such as ibuprofen and aspirin to new, targeted monoclonal antibodies – tend to improve depressive symptoms.[22] One study, carried out in 2020, was able to look at very detailed patient data from eighteen clinical trials and factored in patients' physical improvements. These trials had used a range of new anti-inflammatory drugs designed to target the inflammation behind various autoimmune diseases. Two drugs in particular demonstrated an improvement in low mood, regardless of improvement in physical symptoms. These were both antibody drugs that specifically targeted pro-inflammatory cytokines: sirukumab relieves rheumatoid arthritis by targeting IL-6, and ustekinumab treats psoriasis by blocking the cytokines for IL-12 and IL-23. Here we have data to back up Iain McInnes's observations in his arthritis clinic: anti-inflammatory medication treating the psychological symptoms of depression, regardless of the state of physical symptoms. This doesn't just help to strengthen an argument for inflammation causing some forms of depression; it offers hope of a treatment.

Yet, while this evidence is compelling, it is not convincing on its own. To confirm our suspicions, we need to triangulate data from many different types of research – one of which is epidemiology, the study of patterns of disease across populations. A scientific field little known to the general public before the Covid-19 pandemic, its dependence on complex statistics strikes fear into the heart of many a medical student, and few doctors end up in epidemiology as a specialty. Thankfully, I knew just the person to ask.

Golam Khandaker is Professor of Psychiatry at Bristol Medical School. As a young psychiatrist in the early 2000s, he used both his medical experience and his Master's degree in Epidemiology to look for patterns among those living with depression: 'At the time there was some evidence that there was an association between depression and raised inflammatory markers in the blood. But was it causal? These studies just looked at a population at one point in time, and there was little way of knowing whether inflammation caused depression or whether it was the other way around; the chicken-and-egg problem.' Khandaker was so intrigued by this question that he pursued a PhD in 2008 that led to an academic career exploring the relationship between inflammation and mental health. In the early 2010s he began to investigate the results of a forward-thinking population study that might just begin to answer his question: the ambitious ALSPAC study.

In the first few years of the 1990s, researchers at the University of Bristol recruited around 14,000 pregnant women into the Avon Longitudinal Study of Parents and Children (ALSPAC) study, and have since been monitoring them, their children and partners over the past three decades. Khandaker found that among this treasure trove of physical, mental and social data, 'they had measured the inflammatory markers CRP and IL-6 in around 5,000 children when they were aged nine. This meant that we could look at these children when they were eighteen and see if inflammation in childhood predicted depression later on. And indeed it did: kids with higher inflammatory markers in their blood aged nine were more likely to have depression when they were eighteen.'[23] This was still the case when Khandaker's team statistically removed the effects of potential confounding factors such as sex, age, ethnicity, body mass index, past psychological problems and social class. 'But this was not enough,' Khandaker says, with the angst of someone unsatisfied with a half-truth. He wanted to dig down beneath all of the layers of muddle. 'We had accounted for many confounders, but I'm sure we can't account for other factors we simply do not know about. I decided to turn to "Mendelian randomization".'

In theory, the best way to find out whether inflammation causes depression would be a randomized trial, in which individuals would be randomly allocated to either somehow being given chronic peripheral inflammation or being given a placebo. Both groups would then be monitored for years to see whether individuals in the inflamed group were more likely to develop depression. Clearly this is ethically unconscionable, but clever geneticists have realized that we have all been entered into a natural randomized trial: at conception. Gregor Mendel is now known as the father of genetics, but to those who knew him back in the nineteenth century – in an obscure monastery in what is now the Czech Republic – he was just another monk with a penchant for gardening. Through a series of ingenious experiments in which he crossbred pea plants and noted their characteristics (such as height, pod shape and seed colour) across successive generations, he discovered that specific, discrete traits were inherited according to a set of rules. One of these 'Mendelian laws' is the Law of Segregation. Traits (what we now call genetic characteristics) have alternate forms (now called genetic variants, or alleles). Let's use hair colour as an example. Each sperm in a male and each egg in a female carries just one of these hair-colour variants, which has been allocated to it at random. When the sperm and egg fuse at conception, these variants pair up, creating a new combination. 'The gene that gives me black hair – I've inherited that independently from the gene that gives me my brown eye colour,' Khandaker explains, 'and each one is a random combination from my mother and father. In the same way, there are different genes that encode for an individual's baseline levels of inflammatory markers – and these variants cannot be the consequence of disease or any other confounding factor.' There was in fact an ethical way for Khandaker to run his perfect experiment: in 2021 his team used this technique and discovered that levels of the inflammatory marker CRP appeared to be inherited, with an increased risk of developing depression.[24] Supporting the concept of an inflamed depression subtype, this genetic predisposition was associated with depression characterized by sickness behaviour symptoms: anhedonia, fatigue and changes in appetite. This doesn't cover every case of depression, but it looks like we can now account for one kind, at least.

A holy grail?

Our evidence is mounting: we have mechanisms linking inflammation and depression, and this plays out in large population studies. This leads us to our final, practical question: do anti-inflammatory therapies effectively treat people with inflamed depression?

The most tried-and-tested method for finding out if a drug is genuinely more effective than another treatment (whether that be placebo, standard care or the best available drug) is the double-blind randomized control trial (RCT). This is a bit of a mouthful, but it's fairly logical. Participants are 'randomized' to being given either the new drug or, let's say, a placebo, in a way that controls for potential confounding factors such as age, sex and ethnicity. Participants are statistically 'controlled' so that the two groups are as similar in baseline characteristics as is possible. 'Double-blind' means that both the participants and the researchers do not know whether the treatment being given is the drug being tested or a placebo.

The results of the first clinical trial of one of the new generation of anti-inflammatory drugs were published in 2013. Eminent American academic psychiatrists Charles Raison and Andrew Miller led a study at Emory University in Atlanta, Georgia, in which patients with depression were given either a drug called infliximab or a placebo.[25] Infliximab is a biologic drug that targets the pro-inflammatory cytokine TNF-α and is used in a number of autoimmune conditions. But the results were disappointing: no difference in depression-score reduction between infliximab and the placebo. But when they analysed the data further, they saw that a subgroup of patients did respond well to the medication: those who tended to have a mildly raised background level of inflammatory markers. Specifically, these were patients who had a raised baseline level of C-reactive protein (CRP) – a generic marker of inflammation in the body. A later, larger study assessed the anti-IL-6 drug sirukumab on patients who had not responded to conventional antidepressants and who had raised baseline levels of CRP.[26] In this group, the drug improved only the depressive symptom of anhedonia (inability to experience pleasure), though in patients with a very high background CRP, it helped relieve depression better than the placebo. There are positives to take from this mixed bag. It seems that a calming of inflammation in the body (we know that these antibody drugs do not primarily access or target the brain) can treat depression in those with some level of peripheral inflammation.

Alongside testing medications that calm down peripheral inflammation, researchers are also trying to treat depression by directly calming down inflammation in the brain – neuroinflammation. The drug of choice is an old, cheap antibiotic that was formerly used for the treatment of acne. Minocycline has been identified as a potential treatment for inflamed depression for a number of reasons. The drug is a small molecule that can cross the blood-brain barrier, helping to stop microglia from becoming pro-inflammatory and also dampening down the inflammatory kynurenine pathway in the brain.[27] Recent small trials of minocycline have painted a similar picture to newer biologic drugs: they are not effective antidepressants when given to a general population of people with depression,[28] but seem to show efficacy in patients with mild peripheral inflammation.[29]

So how do we identify those who will benefit from anti-inflammatory treatment? In other words, how are we to define and diagnose inflamed depression? It's clear that inflammation is involved in a minority – albeit a large minority – of cases of depression, so we need a way to identify these cases. For many researchers, the holy grail is a 'biomarker': an objectively measurable marker that is indicative of a particular disease or body state. In 2020, researchers at the University of Cambridge, alongside a consortium of experts from across the United Kingdom, published some exciting findings. They analysed a range of immune cells taken from blood samples from around two hundred patients with depression and three hundred healthy controls. Individuals with depression generally tended to have considerably more peripheral inflammation than healthy controls, but researchers were also able to clearly identify a specific 'inflamed depression' subgroup, comprising around a third of those with depression.[30] Alongside raised CRP and inflammatory cytokines – something already well established from other studies – they also found raised numbers of both innate and adaptive immune cells circulating in the blood of these patients. This subgroup also had more severe depression than others, reflecting other findings showing that those with inflamed depression tend to be those who don't respond to conventional antidepressants.[31] Perhaps we are not too far away from a future in which you present to your GP with symptoms of depression, are given a quick blood test and – if you're identified as having inflamed depression – are offered a cure.

It is clear that those with depression and raised inflammatory markers in the blood tend to have a specific symptom profile compared with others with depression. These patients have symptoms that overlap with sickness behaviour: anhedonia, low energy, low concentration and slowness of mind and body.[32] As Professor Harrison explained to me on my visit to Cardiff, one of the key results of inflammation in the brain is a dysregulation of reward processing: individuals become much less willing to expend effort for a perceived reward. This is something that would be difficult for even a well-trained mental health professional to tease out, let alone a GP, who has to keep up a vast knowledge of every single body system. Perhaps a well-validated questionnaire or a computer-based reward task similar to the one I carried out in Cardiff, combined with blood results, would aid clinicians in reliably diagnosing inflamed depression. Maybe neuroimaging will also one day help to identify the relevant subtypes of the condition, although reliability and cost mean that this reality is still some way off.

No chicken, no egg

I am in danger of suggesting that the only way out of depression is medication. For many doctors, researchers and sufferers, the possibility of a new, effective anti-inflammatory antidepressant is tantalizing. The idea of a silver-bullet pill is attractive to many as it constricts the complexity of human biology and psychology into half an inch of compressed powder. I have to admit, this thinking may have influenced my trip to Amsterdam to meet another expert.

Professor Brenda Penninx is a renowned psychiatric epidemiologist based in Amsterdam's University Medical Centers (UMC), set in a leafy campus on the outskirts of the city. After I've given a talk in the nearby city of Utrecht, I can't miss the opportunity to speak with her. When I meet Penninx, she's working on the INFLAMED trial, which is assessing whether the anti-inflammatory drug celecoxib can help treat depression in those with both raised inflammatory markers and the symptoms profile of inflamed depression.[33] I begin by asking about her role in identifying an inflamed subtype of depression. 'When I first studied the symptoms we use to diagnose and classify depression, I was confused. One symptom is "appetite disturbance", but for some people that means eating less, for others it means eating more. And it's the same with sleep: some people suffer with insomnia, others with hypersomnia. I don't know of many other health conditions where there are symptoms that can go in opposite directions! It is clear that the underlying disease process behind depression is very diverse.' As with Golam Khandaker in Bristol, much of Penninx's research involves looking at disease patterns in large populations, including the Netherlands Study of Depression and Anxiety (NESDA), which she leads. One of Penninx's first investigations was to find out which groups of symptoms of depression tend to cluster together. One group that was clearly evident was typical of what we would classically call depression: insomnia, loss of appetite, low mood and high suicidality.[34] 'But also we noticed another group,' Penninx tells me. 'Roughly a quarter of people with depression have a symptom pattern that is more "neuro-vegetative": hypersomnia, hyperphagia (eating more), energy loss, fatigue and slowed movements. They don't differ in the severity of depression . . . they just have a very different pattern of symptoms.'

Do these symptoms sound familiar? This clustering of behaviours and feelings maps neatly on to sickness behaviour and the 'inflamed depression' subgroup found by other researchers. Penninx's team, however, have also added to this model. Alongside evidence of chronic inflammation in this group, they also found a number of disturbances in metabolism – the conversion of food to energy. Their inflamed depression group contained high numbers of individuals with insulin resistance (a loss of regulation of blood sugar levels), leptin resistance (the reduced effectiveness of a natural appetite suppressant), and an imbalance of lipids and cholesterol in the blood. These abnormalities can be caused by chronic inflammation, and they can also drive up inflammation itself. No wonder this group has higher levels of 'metabolic syndrome' – a combination of obesity, diabetes and high blood pressure. To reflect this, Penninx named this condition 'immunometabolic depression'.[35] Regardless of whether we call this large subgroup of depression 'inflamed' or 'immunometabolic', it is clear that – in Penninx's words – 'chronic inflammation is the substrate that connects depression with these so-called "physical" diseases'. In other words, chronic inflammation is the soil from which the weeds and thorns of modern diseases sprout: diabetes, heart disease, inflamed depression, autoimmune conditions and – as we will see later – dementia. These diseases are the result of a body's defence system that is chronically overactive. The link with metabolic diseases is also interesting as it is clear that when the defence system is unbalanced, it is likely to take the body's energy system down with it. If a country spends too much of its budget on its military, all other areas of national life inevitably suffer.

I posed a question to Penninx that I have been asking throughout this chapter: does inflammation cause depression or is it just associated with it? After a short pause, Penninx's answer was contemplative: 'There is a lot of debate as to whether depression is caused by inflammation, or whether it is the other way round. My opinion is that there is no chicken and there is no egg. What we have found is evidence for directionality both ways: inflammation causing depression, and depression causing inflammation.[36] It is ultimately a circle – or rather, a vicious cycle. The question is: how can we best break this cycle?' Penninx is looking at many potential solutions. Within her own large research group and across collaborations, she is investigating interventions that target diet, exercise, stress and childhood trauma. She also believes that, while lifestyle and psychological interventions are clearly important, anti-inflammatory medications may play a role in treating inflammation-related depression.

As I left Penninx's office and walked out into the soft light of a late-spring afternoon, I looked around the campus of Amsterdam UMC. Commuters pedalled by on wide cycle lanes. Two researchers sat on a bench facing a reed-lined canal, absorbed in conversation. Others wandered down footpaths that led to the many areas of wooded green space that dotted the area – parks and woods that urban planners had clearly prioritized. The following day I eschewed the main tourist sites of Amsterdam to visit Micropia, the world's only microbe museum. As I wondered at the ecosystems that reside in and on us – and how modern Western diets have all but desertified our microbiomes – Penninx's focus on lifestyle medicine began to make more sense. There is much that is good in modern, developed society, but there is also much that stokes inflammation's fire. From low-diversity diets to a lack of regular movement, from disrupted sleep to constant, slow-drip stress, it is no wonder that the most common diseases of the modern world have their root in inflammation.

Where now?

This chapter only sketches out the multifaceted relationship between inflammation and depression. But a clear picture is emerging. For many people living with depression, inflammation is the driver. For some, inflammation has triggered depression. For others, inflammation may have resulted from depression, or from a separate source completely, but now accelerates the downward spiral. The concept of 'inflamed depression' doesn't fit into the neat compartments of 'mind' or 'body' into which we try to awkwardly squeeze diseases. Depression is a whole-body disorder. In fact, all diseases – whether we like to label them under the moniker of mental or physical – are whole-body disorders. We saw in the last chapter that schizophrenia can bubble up from an abnormal immune system – itself a result of past infections, stress and genetic predispositions.[37] When dealing with disease, we must be inclusive of mind and body. This requires making peace with uncertainty, but it is more rewarding and, ultimately, closer to the truth.

Perhaps the best way of drawing this chapter together is to use analogy. An emerging theme throughout the chapter has been one of vicious circles, so let's start with a wheel. Imagine that your defence system – comprised of mind, body (with a focus on the immune system) and microbiome – is a wheel of a bicycle that you are cycling along the road of life. Roughly a third of this is your mind, a third your body and a third microbiome, but it's difficult (if not impossible) to tell where each part starts and the next ends. They all influence each other to the point of being indistinguishable. There are some things – completely out of your control – that might start to knock the wheel out of balance. It could partly be the makeup of the bike – your genetics. It might also be the uneven road behind you – your socioeconomic environment, past infections, and formative experiences of abuse or trauma. Other obstacles lie in wait that could further knock the wheel out of balance: lack of sleep, insufficient exercise, poor diet and stress. These factors might primarily affect the 'mind' part of the wheel, the 'body' part, the 'microbiome', or even all three, but the end result is the same: the damaging friction of chronic inflammation.

Later, in the final part of the book, we will explore a number of evidence-based and accessible ways of rebalancing the wheel of your defence system. But we have some fascinating pieces of the puzzle to locate and slot into place before we get there. In the past two chapters we've mainly looked at how the immune system can influence the mind, from rogue antibodies causing psychosis to inflamed depression. Now it's time to look from the other direction: how the mind can powerfully affect the immune system.

8

Inflammatory Thoughts

How your mind affects immunity

Nothing but man of all envenomed things,
doth work upon itself, with inborn stings.

JOHN DONNE,
ELEGY ON THE LADY MARKHAM, 1609

The stress of shopping

Imagine that it is an unremarkable, overcast Saturday morning. You've had a shower and are getting dressed for your primary tasks of the morning: hunting and gathering. You have a family to feed for the next week, so, armed with an assortment of bags-for-life and a shopping list, you head to the front door. As you step out on to a quiet street, something in the foreground catches your eye. Half-hidden behind a neighbour's bin, with only its camel-coloured hind legs in view, appears to be a large, unattended dog. Out of curiosity, and with the intention of reuniting the unfamiliar pet with its owners, you approach the bin. And then you see it: crouched down in a coil of lean, muscular limbs is a sabre-toothed tiger. A pair of amber eyes fix directly on you. Beads of saliva run down oversized fangs, dripping on to the pavement. It seems that you are both in search of lunch.

Time stops. Your whole being sharpens. Adrenaline courses through your body, unlocking your liver's stores of glucose and diverting it to your muscles. Your heart rate and blood pressure sky-rocket as blood is rapidly redistributed towards those muscles. Hairs stand on end. Your mouth dries up and your bowels prepare to empty; digestion can wait. Before you've even realized it, your body is primed for swift, violent action. Your mind is also sharpened: you become alert and hyper-focused on your goal, whether that be darting back into the house or preparing to grapple with the beast. And one layer below even your conscious awareness, another remarkable thing happens: your immune system is activated. An alarm is sounded and the barracks where your immune army lives – your bone marrow, spleen, lymph nodes – are emptied of their soldiers. Immune cells travel through the bloodstream to barrier surfaces such as the skin, preparing to meet a potential infectious invader. A single thought – the brain's recognition of a mortal threat – has not only changed the physiology of much of your body, it has also changed your immune system.

Stress is good! In the short term, at least. In this 'fight, flight or freeze' response, you become temporarily superhuman. This is total war: your whole defence system is mobilized into action. Looking at it this way, it should be no surprise that something 'psychological' – a mental recognition of a threat – results in numerous powerful 'physical' responses. It should also therefore be no surprise that the mind can rapidly influence the immune system: it is all part of the same response. More simply put, it is a whole-organism reaction. So far in this book, we have mainly explored how your immune system can influence your mind. But the mental and immune elements of our defence system should have equal parity in our discussion, and in this chapter we will see how our mind can influence our immune system, for good or ill.

The stress response is perhaps sickness behaviour's mirror image. Sickness behaviour is the alteration of the mind in response to a dangerous microorganism being recognized by the immune system. Stress is the alteration of the body in response to a dangerous situation being recognized by the mind. They might seem like opposites, but they both represent an activation of a unified defence system. When we realize that every element of our being is fine-tuned to detect and respond to threats, the mind-body debate becomes immaterial.

Now imagine a more realistic scenario: as you leave the house you notice that a neighbour's car is blocking your driveway; or you discover your bicycle tyre is flat; or you're plagued by thoughts of how you are struggling to both feed your family and pay the rent; or your mind is occupied by the headlines you heard on the radio over breakfast: climate change, economic uncertainty, online trolling, the crisis of caring for the elderly. The human organism has not been designed and refined to flourish under the pressures of our modern world. It has adapted to rapidly respond to often-fanged and sometimes-furry threats. These threats may have been serious, but they were also encountered infrequently. Today, we live instead in an unnatural environment of slow-drip, persistent stressors. This has profound implications for our defence system and, ultimately, our health.

Before we look at stress, we need to look at other ways in which our thoughts can influence our immune system. From reading the previous two chapters, you might get the impression – understandably, but mistakenly – that our brain and mind are at the whims of the inflammatory substances produced by our immune system. But now we are going to look at the other side of the defence system. Let's explore the true power of the mind.

Mind over microbes

In Chapter 2, while we were dissecting the multiple pathways between brain and immune system – between mind and body – we were introduced to a remarkable study published in late 2021. A team led by immunologist Professor Asya Rolls at the Israel Institute of Technology used new research techniques to pinpoint the exact neurons in a mouse brain that were activated during experimental bowel inflammation.[1] This is impressive in and of itself, but their next move was the *pièce de résistance*. Well after the mouse had recovered from the bowel inflammation, they used a technique called chemogenetics to reactivate those same neurons, nestled in the insula of the brain. This is the same region that, nearly a decade earlier, Professor Neil Harrison had identified as being a key modulator of sickness behaviour. In the complete absence of infection or irritants, this brain stimulation in mice brought about the exact same inflammatory state in their bowels. What we have here is a clear picture of a specific area of the brain encoding a specific memory of prior inflammation, and the activation of this memory being enough to reinflame the body.

It is still uncertain how the remarkable specificity of the reactivated immune response is achieved. Asya Rolls wonders whether a memory is not just encoded in the brain, but is also stored at the tissue site in the form of memory cells of the adaptive immune system.[2] Perhaps it is no surprise that peripheral immune cells can interpret specific messages coming from the brain; we know that the immune and nervous systems speak the same language. And perhaps this is less surprising still when we take a step back and remind ourselves that the immune system and nervous system are part of an integrated supersystem. Regardless, what would have been considered science fiction a few years ago has been shown to be very real indeed. Your brain constantly monitors your immune system, but your immune system is also listening to your thoughts.

Cutting-edge studies like this are beginning to provide concrete evidence for something we feel intuitively. We all have a diffuse, intangible sense that our mind can influence our immune system; that mental wellbeing affects immunity. Experimental evidence for this has existed for well over a century. In September 1885, Baltimore physician Dr John Nolan Mackenzie exposed one of his patients, a thirty-two-year-old woman who suffered from severe seasonal pollen allergy, to 'an artificial rose of such exquisite workmanship that it presented a perfect counterfeit of the original'.[3] Although she had been feeling 'unusually well' when she entered the clinic room that day, exposure to the fake flower activated a powerful inflammatory response: 'In the space of five minutes she was suffering from a severe coryza ... the right nostril was completely obstructed by the swollen, reddened, irritable turbinated structures.' Dr Mackenzie's notes on the case are a fascinating insight into someone trained as a mind-body dualist beginning to see behind the curtain. 'We should distinguish carefully between a disease ... subject to recognized pathological law, and a mere perversion of the perceptive faculty, although the latter may occasionally act as an exciting influence in the production ... of the former.' In other words, Dr Mackenzie was beginning to accept evidence that the mind can exert power over the body.

The medical world at the time kept mind and body as far from each other as possible, so almost no research exploring the mind's influence on the immune system was carried out for another century. Then, in the 1970s, researchers accidentally found that if you gave rats an immunosuppressant drug paired with saccharin (a substance with a distinctive taste), a dose of saccharin alone given at a later date would be enough to trigger immunosuppression.[4] The rats had become conditioned to expect immunosuppression when exposed to a specific flavour. Plenty of subsequent studies have found that the immune system can be behaviourally conditioned to be either underactive or overactive.[5] The placebo effect – the response our brain makes to the *context* in which a treatment is delivered – has a remarkable influence in the experience of pain – something I have extensively explored in my book *The Painful Truth*.[6] What I did not expect to discover, however, was the extent to which our beliefs, predictions and expectations influence our immune system. Mere expectation – conscious or unconscious – can powerfully change your physiology.

Why should your mind have such sway over immunity? It's because your defence system – not just your brain – is constantly striving to reduce uncertainty and predict potential threats. The mobilization of your immune system in the face of mortal threats, such as a pair of sabre-teeth, is fairly sensible – a violent encounter is likely to result in a breach of your skin's barrier, and with it a host of nasty bugs. But research is revealing that it can take a lot less than extreme stress to alter your immune response. As we've seen before, merely looking at the faces of sick people can trigger inflammation in the body.[7] Mating also provides a number of routes for pathogenic spread, so it might make sense for the immune system to be in a state of heightened readiness. In 2022, a team from the University of Tokyo found that male mice experienced raised levels of the immune cell stimulating cytokine IL-2 when exposed to females. They used new techniques, similar to those employed by Asya Rolls's researchers, to find that these immune changes were caused by dopamine synthesizing neurons in the brain's reward circuits.[8]

Your immune system also pre-emptively readies itself for battle during a specific mood state: anger. It appears that bouts of anger are associated with high levels of inflammatory blood markers in the short term, and chronic inflammation in the long.[9] A short burst of inflammation during anger makes sense from an adaptive standpoint: anger often precedes violence, which often precedes a break in your skin and the introduction of microbial foes. Anger is not simply an emotion; it is a state of our defence system. Our language of anger is one of fire: someone might have a 'fiery temperament', be 'incandescent with rage', become 'incensed' at a decision, or simply 'inflamed'. Perhaps this language bubbles up from an implicit knowledge that we have a silent fire within. Chronic anger is associated with terrible health outcomes; one of the links is most likely chronic inflammation.

There are a lot of things in life that we cannot control, but we do have some control as to how we react to these things. Learning emotional regulation – whether informally or with the help of a clinician – is not simply good for your mental health; it balances your defence system, which benefits all health. It follows that different mood states may be associated with changes in the immune state. This is an underexplored area of science, and accurately measuring the relationship between mood and physiology is difficult, but there is emerging evidence. A 2018 study measured blood inflammatory markers in participants at five time points throughout the day for two weeks, and found that periods of negative emotions were associated with increased inflammation.[10] Other studies have found a link between positive mood and reduced inflammation.[11]

Well before we developed the science to explore how mind and mindset affects immunity, history is littered with stories offering tantalizing glimpses into this relationship. One that struck me was in *Man's Search for Meaning*, the classic autobiography-cum-psychological-thesis of Viktor Frankl, the Viennese doctor who lived through the horror of being interned in – and only just surviving – Nazi concentration camps. He recounts how, in the final months of the war, he was entrusted with the medical care of prisoners suffering with typhus in Dachau camp. Likely spread between prisoners by bacteria dwelling in body lice, typhus fever brought about terrible flu-like symptoms and a widespread body rash. As Frankl had no access to antibiotics, he estimates that around half of his patients died. As rumours of the Allies' arrival in Germany – and with it the prospect of liberation – spread around the camp, one of Frankl's emaciated patients became aglow with hope. He was convinced, by a dream no less, that the camp would be liberated on a specific day: 30 March. But as the fated day approached, it was clear that the Allies were still a long way away. On 29 March, Frankl recorded that the man 'suddenly became ill and ran a high temperature':

On March thirtieth, the day his prophecy told him that the war and suffering would be over for him, he became delirious and lost consciousness. On March 31st, he was dead. To all outward appearance, he had died of typhus. Those who know how close the connection is between the state of mind of a man and the state of immunity of his body would understand that sudden loss of hope and courage can have a deadly effect. The ultimate cause of my friend's death was that liberation did not come, and he was severely disappointed. This suddenly lowered his body's resistance against the latent typhus infection.

I do not want to overstate the case. A positive attitude alone does not cure autoimmune disease; self-affirmations do not shrink tumours. There are many who hyperbolize the power of the mind to affect physical health, implicitly – and sometimes explicitly – saying that we can simply smile our way to wellness, casting moral aspersions over those who can't seem to get out of the pit of chronic illness. But at the same time I believe that – particularly in a country as sceptical as Britain – far more people understate the power of the mind to influence physical health than overstate. Dualistic Western medicine has painted the psychological – from mood states to deep-seated cultural beliefs – as ephemeral. This is because, until fairly recently, it was impossible to measure the physical processes in the brain that constitute mental activity. For too long, doctors have given patients the impression that if there is nothing they can physically measure in the body, the cause of their suffering is imaginary. The implication is that the patient's personality – their weakness of character – is to blame for medically unexplained symptoms. It is also implied that a psychological or social intervention is not 'real' medicine. But to a greater or lesser extent, the mind plays a part in every disease. The weighting between mind and body is different in each case – and often it is impossible to see where 'mind' ends and 'body' begins. That's why I invite you to look at health and disease – particularly long-term, chronic disease – through the lens of the defence system: mind and body united, each influencing the other. With that in mind, let's take a look at stress.

Acute stress

'Too well known and too little understood' is how Hans Selye, the Hungarian-Canadian doctor considered by many to be the father of stress research, summed up his field of study. So let's define it: stress is the psychological perception of an environmental threat, and our body's physical response to it. It is a word that spans the awkward boundary that separates mind and body. This is because it represents a process that cannot be simply categorized as only psychological or physical. The mind-body mêlée of stress can make it conceptually difficult to grasp, for our dualistic society at least. To get round this problem, I like to redefine stress as a mobilization of your body's defence system in response to a perceived threat.

Within a split-second of your gaze fixing on the sabre-toothed tiger, your brain has decided that it is a mortal threat. Before you are even aware of it, your brain has combined the visual data emanating from the carnivorous cat with your prior concepts of what a large-fanged feline represents. A key brain area that is activated in this process is the amygdala. As we saw in the last chapter, this 'threat detector' is often hypersensitive in people with overactive defence systems. Signals are sent to the command centre of the brain, called the hypothalamus, which then communicates this danger to the body through two main routes. The first is the nervous system, specifically the unconsciously activated 'sympathetic nervous system'. This system is terribly named, as it is anything but sympathetic to your body. Within a few seconds, sympathetic nerves activate organs throughout the body to stimulate the fight-or-flight response: increasing heart rate and blood pressure, dilating pupils, emptying bowels, secreting sweat and inhibiting the excitability of genitalia. This is a short-term survival response; rest, sex and digestion can wait. Alongside the firing of nerves throughout the body, the brain also activates the endocrine system, the two primary hormones involved in the stress response being adrenaline and cortisol. This coordinated nerve activation and hormone release is powerful and hugely energy-consuming, so the very hormones that carry out the generation of the stress response also hinder the production of more hormones. These intricate 'negative feedback loops' stop the organism spiralling into an uncontrolled, unremitting stress response. Once the threat has passed, the yin to the sympathetic nervous system's yang can be activated to restore restful balance: the 'parasympathetic nervous system'.

Short-term (better known as acute) stress powerfully affects the immune system. We've all experienced a stress response while giving a presentation or a speech: dry mouth, pounding heart, clammy hands. But what is utterly curious is what happens to the immune system: a hormone called noradrenaline stimulates inflammatory pathways within certain immune cells, which then begin to secrete inflammatory cytokines.[12] Stress is like a battle horn to the immune system. Immune cells are mobilized out of their barracks (lymph nodes, spleen and bone marrow) into the blood, and finally to the walls of the body, such as the skin.[13] This makes a lot of sense: when an invader is approaching your city, you want your soldiers at the walls. There is even evidence for psychological stress increasing the ability of your immune system to benefit from a vaccine administration. Most vaccines contain an 'adjuvant' ingredient, which is designed to stimulate the skin's immune cells to help the immune system recognize and respond to the vaccine. Adjuvants used in vaccines are generic molecules, such as aluminium salts, that activate the immune system in a non-specific way, but it also transpires that psychological stress has a similar effect, increasing the trafficking of immune cells resident to the site of infection to T-cells in nearby lymph nodes.[14] While fascinating, I can't see many immediate practical applications of this phenomenon in medicine, short of doctors intentionally stressing out a vaccine recipient to produce a stronger immune response.

We are also starting to see evidence that other elements of an immune response can be triggered by stress. In 2020, a group of scientists from Nagoya, Japan, identified brain circuits that, when activated by psychological stress, induce fever in the body.[15] To those of us raised in a dualistic society, these connections sound like science fiction. But when we remember that psychology is biology, and stress is a defence system-wide reaction, it should be no surprise that thoughts can change the state of our immune system.

A considerable body of literature now shows that short-term stress activates the immune system, preparing it for a breach of the body's outer defences.[16] But new research suggests that the process is more nuanced than this. In 2022 a team based at Harvard and Icahn School of Medicine at Mount Sinai, New York, employed similar technologies to those used by Asya Rolls to see how the immune systems of mice responded to psychological stress. They found that stress immediately mobilized large numbers of 'neutrophils' – the key foot soldiers of the innate immune system – to leave their barracks in the bone marrow and travel to the peripheral tissues.[17] Neutrophils are immune cells equipped for fighting bacteria by either engulfing them, spraying them with antimicrobial molecules or entrapping them in a net of proteins. But the team also found that while short-term stress temporarily increased antibacterial immunity, this reduces antiviral immunity. Why would this be the case? The most likely explanation is that stress is the result of the defence system's prediction that a bacterial threat – rather than a viral one – is imminent. This makes sense when you consider the lives of our ancestors: stress usually preceded violence, and physical injuries are open doors for pathogenic bacteria.

It has long been established that cortisol is released during stress, and this steroid hormone is a powerful immunosuppressant. The release of cortisol during stress is part of a feedback loop that makes sure inflammation doesn't get out of hand. And so, when you perceive a stressor, one invisible hand activates the immune system while another begins to set the wheels of immunosuppression in motion. This balancing act is beautifully complex, and another layer of complexity lies in the fact that it can vary widely between people and between different stressors.[18] In short, just as your immune system recruits your mind during an infection, your mind can recruit your immune system during stress.

Chronic stress

Stress is the defence system preparing both mind and body for violent impact. As with sickness behaviour, the stress response is a symphony of body and mind; like Beethoven's Ninth, it is as muscular as it is delicate. This organism-wide reaction has consequences for mind and body, which should ideally resolve once the threat has passed. As we touched on in the previous chapter, a defence system that stays active once the threat has passed, or one that is persistently exposed to stressors, wreaks collateral damage. While we should be thankful that the prospect of violent death is uncommon, we live in a world of persistent, slow-drip stress – whether this means waiting for an email that decides our employability for the foreseeable future or worrying about rising energy bills. What these threats lack in life-threatening severity, they make up for in persistence. When you are constantly in a state of defence, you can't rest, heal and build. There is a mismatch between the modern world that we inhabit and the world in which our defence systems adapted to survive.

The effects of chronic stress on the immune system are many and varied. As stress creates an antibacterial environment, antiviral immunity is put on hold, meaning you are much more vulnerable to viral infections during periods of chronic stress. Chronic stress can suppress the effect of 'natural killer cells', remarkable immune cells that not only identify other cells that have been infected with a virus, but can also detect and destroy cancer cells. Natural-killer-cell suppression by chronic stress could therefore be a reason why stress is associated with cancer progression and recurrence.[19] By depleting the immune army, chronic stress can also release pathogens from their previously secure prisons. I had the miserable privilege of experiencing this for myself as I prepared for medical school finals. It was a week before exam season, and I was holed up in a small student flat surrounded by well-thumbed textbooks and Post-it notes spoiled by indecipherable handwriting. One morning, after emerging from a much-needed shower, I glanced in the mirror and saw a sudden eruption of little bumps and vesicles (blisters less than 1 cm in diameter) on a small strip of skin on the right-hand side of my back, just below the shoulder blade: it was shingles. This often agonizing condition (I couldn't sleep for a week due to the pain) is caused by the varicella-zoster virus – the same culprit behind chickenpox. This virus is particularly interesting as it is a member of the herpes virus family, one of the only groups of viruses to show latency. Latency is the ability of a virus to lie dormant within human cells after the initial infection, often for years, without replicating. Varicella-zoster tends to infect humans in infancy, causing chickenpox, before sneaking off to hide in a specific nerve on one side of the body. This is why shingles – the reactivation of varicella-zoster – is limited to a specific patch of skin served by one nerve. Reactivation tends to occur when there are significant disruptions to the body's defence system, whether this be infection by another pathogen, immunosuppression by a drug or virus or, indeed, psychological stress. A more common example of a herpes virus reactivating is that of the herpes simplex virus, bringing about the all-too-visible cold sore.

Not only does chronic stress bring out more antibacterial immune cells from the bone marrow and into your circulation, but these cells tend to operate in a more pro-inflammatory state, too.[20] But if stress hormones are repeatedly pumped throughout the body, these same immune cells eventually become less responsive to their message. The result is the worst of both worlds: a defence system that is chronically inflamed yet unable to respond to genuine threats. It's like having a faulty smoke alarm that keeps going off: it causes you persistent grief yet ironically makes you less likely to respond quickly if your home is actually on fire.

There is also evidence for chronic stress bringing about pro-inflammatory changes in the brain. Stress can 'prime' microglia, which become even more inflammatory when triggered by other stressors.[21] There is a throng of studies in both mice and men that demonstrate a clear link between chronic stress and altered immune function. There is also an emerging body of research on the negative effects of chronic psychosocial stress on the non-human element of your defence system: the gut microbiome.[22] The fruit of this imbalanced defence system is our old enemy: chronic inflammation. As we will explore more later, in Chapter 10, the end result of this is not pretty: life-threatening heart disease, life-ruining mental health conditions and life-stealing dementia.

Given that persistent stress can powerfully affect the immune system, it will probably come as no surprise that stress is closely linked to allergies and autoimmune disease. Even short episodes of stress often precede allergy flare-ups, and chronic stress can result in more frequent flares.[23] One six-month period of my time spent as a junior doctor – the word stressful would not begin to touch it – resulted in the longest and worst flare-up of my eczema that I have ever experienced. It is well known that stress exacerbates auto-immune disease, but there is also evidence to suggest that periods of severe stress – such as post-traumatic stress disorder (PTSD) – can increase the risk of developing autoimmunity in the first place.[24]

PTSD is one example of how trauma not only affects the health of our mind, but also of our body. There is an undeniable link between past traumatic experiences – particularly early-life trauma – and chronic disorders of mind and body. A fascinating line of evidence suggests that the link may be inflammation. When I became a lab rat and underwent my brief inflammatory challenge in Cardiff, in the previous chapter, the team were exploring how inflammation impacts the healthy brain's processing of rewards and risks. It turns out that severe stress in children and adolescents not only brings about an inflammatory response, but it changes the way the brain processes rewards.[25] This could mark the beginning of a self-perpetuating cascade of unbalanced mind and body. Trauma and stress imprint themselves on the body in the form of inflamma-tion, which brings with it psychological distress, which creates more inflammation . . . ad infinitum. Interestingly, there also seems to be a link between childhood trauma and the 'inflamed depression' sub-group we explored in the last chapter. The severity of traumatic events in childhood – also known as adverse childhood experiences (ACEs) – appears to correspond to a lack of response to conven-tional antidepressants. It is also associated with something called glucocorticoid resistance: a combination of inflammation and per-sistently high levels of cortisol.[26] A wealth of evidence shows that traumatic experiences can bring about pro-inflammatory changes in both the brain and body.[27] From a defence system perspective, this probably shouldn't surprise us. An overwhelming psychological shock activates defence circuits in the brain and the immune system, which can feed into each other in an ever-widening spiral of damage. There is a fine line between self-defence and self-destruction. Trauma – along with everything our society likes to pigeonhole as 'psychological' – is a whole-body condition.

Psychosocial stress, however, goes well beyond trauma. What if I told you that there was a disease that can't be found in medical text-books, yet is worse for one's physical health than smoking, causes depression and suicide, is contagious and affects a large and growing proportion of society? This condition is called loneliness. It is counter-intuitive to view the state of being alone as stressful or inflammatory, but humans are social creatures, and your defence system sees prolonged isolation and social rejection as threats to survival. Research is revealing that loneliness is associated with a low-grade inflammation, and with a defence system that becomes over-inflammatory when exposed to stressors when compared with one that gets regular social interaction.[28]

Another significant risk factor for stress and its complications is low socioeconomic status, and this is not simply explained by the dietary results of poverty. Large cohort studies that followed the health status of British civil servants and Finnish industrial workers found that both the lack of a sense of control over one's work, as well as unpredictability in the workplace, predicted adverse health outcomes in response to stress.[29] The perception of being at the bottom of a hierarchy – regardless of workload – is stressful in and of itself. This touches on a profound truth: the social, as well as the psychological, manifests itself in biology. Repeated and sustained oppression – from workplace bullying to systemic racism – can hyper-sensitize a victim's immune system in brain and body and directly contribute to almost every form of chronic disease.[30] Poverty, loneli-ness, trauma and discrimination are not abstract concepts or states of mind; they are etched into our physiology.

We already know that the immune system can easily influence

the mind, but now it is also clear that the mind is a powerful immunological force in its own right. Just as our immune system influences our thoughts and emotions, our thoughts can directly change the state of our immune system. It is a series of never-ending feedback cycles. The environment acts upon you and you act upon the environment. You are a loop. When your defence system is healthy and balanced, this virtuous circle can bear the load of short-term threats, recalibrating once the danger has passed. But, for one reason or many, this system can lose balance and, with a mind inflaming the body and the body inflaming the mind, this loop becomes a vicious downward spiral. I appreciate that I have spent a lot of time in the negative, but a mechanic needs to know how and why things go wrong before starting the repair job. In the final chapter of this book, we will see that there are many points of the loop at which we can intervene and recalibrate an unbalanced defence system. But before that, let's see if our understanding of the defence system can help us approach one of the most contested areas of medicine.

9

Nobody's Land

The truth is rarely pure and never simple.

OSCAR WILDE,
THE IMPORTANCE OF BEING EARNEST

Bir Tawil is a strange place. In January 1899, administrators of the British Empire decided to finalize a border between the countries of Egypt and Sudan. In colonial fashion, they decided to use the conveniently straight line of the twenty-second parallel of latitude. Unsurprisingly, this new international border arbitrarily tore through local tribes and communities, and in 1902 the British drew a new line to reflect tribal territories. The combination of these two intersecting lines brought about the creation of two new areas: the large, inhabitable coastal area called the Hala'ib Triangle and an empty, landlocked patch of desert, which is Bir Tawil. To keep the desirable Hala'ib, Egypt understandably claimed the original, straight border of 1899 whereas Sudan – with the same goal in mind – claimed the new 1902 line. The bizarre result of this impasse is that, to this day, both countries refuse to claim Bir Tawil, an area of land roughly the size of London, as it would mean ceding a greater prize. This makes Bir Tawil the largest – perhaps the only – piece of land that no country wants. It is *terra nullius*: nobody's land. The only occasional occupants are small groups of migrating tribes.

There is a large *terra nullius* in medicine, too. Millions upon millions of people suffer with long-term conditions that are claimed by no medical specialty. While there is considerable overlap, they tend to fall into one of two main categories. One category is characterized by pain – from the many forms of chronic pain that persist well after an injury has healed, to the whole-body pain of fibromyalgia. The other category is distinguished by persistent, severe fatigue – from myalgic encephalitis/chronic fatigue syndrome (ME/CFS) to post-viral illnesses such as long Covid. Although largely unclaimed by the nation states of medical specialties, they certainly are fought over. A bewildering array of tribes – differing patient groups, scientists, activists, journalists and doctors – have a stake in these 'contested illnesses', but they tend to polarize into two opposing camps.

The 'Minds', as I'll call them, believe that these illnesses are psychological in nature: perhaps a journalist derogatorily dismisses ME/CFS as 'yuppie flu'; someone with a Freudian bent may hypothesize that the mental and neurological symptoms of long Covid are a socially acceptable outlet for psychosocial suffering, repressed anger or childhood trauma; a doctor might try to reassure a patient that, as their tests have come back negative, their condition isn't serious, or doesn't *really* need treating. I remember my first year as a junior doctor, when the words of one consultant became etched on my mind. An IT worker in his late forties had come to A&E with terrible, unremitting long-term back pain that had worsened over the past few months. That morning, he couldn't even get out of bed. The consultant looked over the pristine MRI scan images and at the string of normal blood results, before confidently announcing to the man: 'The good news is, there's nothing physically wrong with you!' As the consultant glided into the next bay to continue his morning ward round, the patient – arched uncomfortably on the hospital bed and drenched in dolorous sweat – was left with two options: either to believe that the pain was 'all in his head' and that he should be able to just think it away, or that it stemmed from some terrible structural cause that the doctors couldn't find.

Those who belong to the tribe at the other end of the spectrum are the 'Bodies'. At the extremes, Bodies believe that the experiential symptoms of chronic diseases – from fatigue to pain to brain fog – are always the brain's response to an ongoing, destructive 'physical' cause, such as a persistent virus or organ damage. If there is brain dysfunction, it is never a 'software' issue of brain circuits, but instead a case of irreversible and irreparable 'hardware' damage to neurons: neurodegeneration.

As you can probably guess, in this short chapter I am not going to take either of these (admittedly caricatured) sides, but instead try to offer a different starting point in approaching seemingly unexplainable diseases. Each of the conditions I explore is nuanced and personal to those living with them; it would require a whole book to do just one of them justice. But, for our purposes, we'll look at the neuropsychiatric symptoms of long Covid, as this condition clearly illustrates the brain-immune connection in post-viral illnesses, is probably familiar to most readers and is a condition of which I have some personal experience.

Soon after the first wave of SARS-CoV2 viruses had crashed on to British shores in March 2020, I was redeployed to a Covid ward. It was a terrifying time. With no treatments available, except for high-flow oxygen and the option (for those who were fit enough) of invasive ventilation – we looked on helplessly as some people recovered, some were rushed to the intensive care unit and others rapidly deteriorated and died. Dressed in wholly inadequate protective equipment – an ill-fitting plastic visor for my face and a bin bag for my body – I inevitably contracted the virus within a couple of weeks. My first symptom, experienced by many others, was a complete loss of smell and taste. Remarkably, this carried on for almost nine months until – either due to my increased persistence at 'smell training' or due to natural recovery – these senses returned.

It was my third Covid infection that truly made me aware of the ominous lingering neurological effects of the virus. It was in the spring of 2022, and I had just recovered from a fairly mild bout of Covid that was making its way through my hospital once again. On the first day I returned to work, I read through the list of outstanding jobs I needed to complete on the ward and left the doctors' office in search of a patient who I had marked down as a priority. But as soon as I let the door of the office shut, I felt lost: who was I meant to be seeing . . . and what for? I had to go back into the office to check. These lapses in short-term memory – combined with an embarrassing difficulty in finding the right words and an inability to concentrate properly – characterized a roughly six-week period of mild cognitive impairment, sometimes termed 'brain fog'. I was lucky that these worrying symptoms resolved within two months, but many aren't as fortunate.

Post-viral illnesses remain poorly understood and have very little consensus in the area of treatment. There is in fact no consensus as to what long Covid actually is, but we can define it as a syndrome of symptoms and health problems that persist beyond Covid's usual recovery period of four weeks. Crucially, long Covid is hugely heterogeneous (a fancy scientific word that essentially means 'diverse') – presenting with any combination of 203 symptoms.[1] As we saw with the variety of shapes that depression can take, this heterogeneity is strongly suggestive of a range of different mechanisms at play. It's easy to latch on to simple answers to explain the persistence of long Covid. The human brain is allergic to chaos, and humans search for meaning more than anything else. It is no wonder so many are drawn towards settling on a one-cause, one-effect, one-treatment explanation for disease.

Most of the symptoms of long Covid affect the brain and how it relates to its body: fatigue, brain fog, anosmia (loss of smell), palpitations, dizziness, low mood, anxiety and joint pain, to name a few. This leads us to a key question. We understand how organ damage resulting from an acute infection that's severe enough to require hospitalization and the use of a ventilator can linger on in the body. But how can a virus that causes, in most people, a mild and transient infection bring about long-term brain changes?

Many researchers have explored Covid-triggered persistent changes to the immune system. One hypothesis that I find particularly interesting posits that Covid infection reactivates viruses that normally lie dormant in most healthy individuals: such as the Epstein-Barr virus (EBV). This herpes virus is a relative of the varicella-zoster virus that, as we saw in the last chapter, reawoke across the skin of my back in the form of painful shingles during a stressful exam period. But instead of lying latent within peripheral nerves, the EBV virus resides within B-cells of the immune system. In the acute, initial phase of infection, EBV causes infectious mononucleosis – also known as glandular fever or 'mono' – before the immune system responds and locks these viruses back in their host cells. But, over time, these sleeper agents can come out of hiding and greatly increase someone's risk of developing various blood cancers, autoimmune diseases and post-viral illnesses. Threats to the immune system – from repeated infections to immunodeficiency – can reactivate sleeper agents like EBV. There is also some evidence to suggest that psychosocial stress can do the same thing, providing one mechanism behind the observation that stress can be a trigger for seemingly unrelated conditions.[2]

It now appears that Covid infection can bring about EBV reactivation, as shown in three studies published in 2022. One Chinese study found that a quarter of those infected with Covid exhibited EBV reactivation, and that this was more common in women.[3] There is also emerging evidence implicating EBV reactivation in long Covid: another study found that EBV particles detected in the blood during the time of acute Covid infection predicted a subsequent development of long Covid.[4] In addition, 2022 saw an American team led by Professor Akiko Iwasaki, a renowned Yale-based immunologist, carry out detailed profiling of the immune cells and molecules of these patients and finding that, compared with healthy controls, their immune systems tended to be skewed towards dealing with a recent reactivation of EBV.[5] These individuals were more likely to harbour antibodies directed against EBV proteins in their blood, and they were also likely to have immune cells producing an inflammatory response to these proteins. This state of chronic inflammation in the periphery can easily bring about the long-term symptoms of sickness behaviour, as we have already explored in depth. Perhaps these particles result from a reactivation of EBV, perhaps they are particles of the SARS-CoV-2 virus that have somehow persisted in reservoirs within the body, or perhaps a combination of both. Or, in some people, perhaps it's neither. We have to appreciate that not everyone with long Covid has evidence of reactivated EBV, and for those who do, we don't yet know whether anti-EBV strategies will work. There is, however, another reason to consider EBV a threat to the brain.

There are few scientific papers that affect me to the extent that I remember exactly when and where I was when I first read them. At around 3 a.m. on 14 January 2022, in a lull between seeing patients on a night shift, I came across an astonishing study. Using detailed data from millions of US military veterans over a twenty-year period, a team from Harvard University found that the EBV infection is responsible for the vast majority of cases of multiple sclerosis (MS).[6] MS is an immune disorder in which the immune system attacks myelin, the insulating sheath that surrounds neurons in the brain. The resulting neurological dysfunction can manifest in almost any way: blurred vision, blindness, muscle weakness, numbness, fatigue, depression, a loss of balance, muscle spasms and incontinence. This study demonstrated that EBV infection confers a thirty-fold increase in one's risk of developing MS. But EBV clearly does not inevitably lead to MS: 90 per cent of the population are infected with this virus, yet only a fraction of 1 per cent of people will develop this neurodegenerative disease. In short, it seems that EBV is necessary but not sufficient on its own to cause MS. It loads the barrel, then something else pulls the trigger. Even if we don't fully know how this insidious virus causes MS, there is every reason to believe that an EBV-targeted treatment or an EBV vaccine – one of which is currently in development[7] – could prevent an awful amount of suffering.

There are a number of other immunological theories as to how the SARS-CoV-2 virus could affect the brain and the mind in the long run. One is that the virus directly invades the brain. This currently looks very unlikely, and while it may happen in rare cases, autopsies have not found evidence of viral material in the brain.[8] Another is autoimmunity – the virus triggering the immune system to continually produce antibodies against healthy tissue. Covid, like many infections, can trigger cases of autoimmune encephalitis – similar to Samantha's case in Chapter 6 – but, again, cases of antibodies directly attacking the brain appear to be very uncommon.[9] What is more likely is the formation of antibodies against other tissues in the body contributing to a generally pro-inflammatory milieu that indirectly contributes to inflammation in the brain.[10]

Another line of research is exploring whether Covid can trigger and propagate inflammatory signals within the brain – neuroinflammation. Studies in mice and humans have shown that even mild Covid can bring about raised inflammatory markers in the brain, a reduction in myelinated (insulated) nerve cells across the brain and a reduction in the production of new neurons (neurogenesis) in the hippocampus.[11] Overall, this neuroinflammation is a recipe for cognitive dysfunction: impairments in memory, attention, learning, processing speed and problem solving. This sounds dramatic – and it is – but the permanence of these changes is debated. My brain fog lasted just over a month, thankfully, but some experience it for much longer. During some cases of severe Covid it appears that the nature of the brain's microglia can start to look like those of Alzheimer's patients.[12]

Indeed, at the time of writing it seems that there is evidence for longer-term increased cognitive dysfunction following Covid infection. In 2022, Dr Maxime Taquet, one of my colleagues at the University of Oxford Department of Psychiatry, carried out an impressive analysis of the anonymous health records of 1.5 million people who have had a diagnosis of Covid infection, comparing these with an equally large population sample of people diagnosed with a different respiratory infection.[13] He found that even two years after a Covid infection, individuals have an increased risk of cognitive deficit (brain fog). Another study, published in early 2023 by a team at Cambridge University, discovered an intriguing link between the mechanisms behind long Covid and inflamed depression.[14] They found that individuals with long Covid tended to have inflammation-induced increases in the kynurenine pathway, which, as we saw in Chapter 7, can result in reductions in serotonin and a greater number of neurotoxic molecules. This might plausibly contribute to low mood, fatigue and pain. While further studies are clearly needed, it appears that the science – from the microscopic, neurobiological level to the population level – suggests that even mild infections can lead to long-term issues.

A curious finding from Professor Iwasaki's study at Yale was that the strongest predictive marker for long Covid was not a specific immune marker, but a low baseline level of the hormone cortisol.[15] Although cortisol rises in short-term stress, there is some evidence to show that long-term disease and stress can bring about low levels of the hormone. This is reliably seen in many other post-viral syndromes. This is not at all to say that post-viral illness is simply the result of stress or burnout. Many triggers – pathogenic or psychological – can bring about a protracted stress response. Perhaps low cortisol should be viewed more broadly as the result of an unbalanced defence system. This leaves us much more vulnerable to the effects of environmental stressors, whether they be physical, psychological or social. This also suggests that having an unbalanced defence system might leave us more vulnerable to getting a post-viral illness, regardless of where the disruption in the defence system began – brain, immune system or microbiome. This theory is supported by the known risk factors identified for developing long Covid: immune disruption (from autoimmunity to EBV reactivation),[16] psychosocial disruption (perceived stress, anxiety and depression),[17] and dysbiosis of the microbiome.[18] These seemingly unrelated factors should not surprise us by now: they all represent a defence system out of kilter.

You could argue that most chronic illness is ultimately a failure of the human organism to resolve a defence response. Understanding this also provides us with a balanced framework when it comes to treatment and recovery. While we should continue to research potential immune-targeted therapies, we should not neglect the other two parts of the defence system. Early evidence suggests that improving the gut microbiome via prebiotics and probiotics could alleviate long Covid symptoms.[19] A large US-based trial – the remarkable results of which were published in June 2023 – found a 41 per cent decrease in development of long Covid in those treated with the diabetes medication metformin after initial infection compared to a placebo.[20] This is particularly interesting considering that metformin has a powerfully positive effect on the gut microbiome.[21] But what we really need, if we want a population-level improvement in defence system resilience, is solid research into the effects of microbiome-focused dietary interventions on the prevention and treatment of post-viral illness. Finally, when we consider the mental aspects of the defence system, there should be absolutely no shame or stigma in using the power of the brain to relieve the symptoms of long Covid. We already know that the state of our mind and nervous system can powerfully influence the immune system and contribute to a hyperactive defence system. A well-targeted psychological therapy with a clued-up, understanding clinician is, ultimately, a powerful biological treatment.

This leads us to a final, mysterious and utterly fascinating piece of the puzzle. In Chapter 4 we met what is perhaps the most exciting theory in neuroscience: predictive processing. Also known as predictive coding, this idea posits that your brain does not passively receive signals from the outside world and inside your body, before decoding them and finally responding. Instead, your brain actively predicts the world – it is an inference machine. This explains why we can be fooled by optical illusions: we see what we *expect* to see. When sensory signals reach the brain, they only reach our perception if the 'prediction error' (the difference between what the brain's model of the world expects and the actual reality of the world) is big enough. The brain seems to be wired in this way in order to save energy: instead of having to compute every bit of sensory data flooding in from the body and the outside world, it only has to react to significant differences between the expected and the actual. All conscious human experience lies in the space in between the brain's model of the world and the actual state of that world. The brain is often right: our very survival relies on its accuracy in recognizing and responding to unexpected states. But we are beginning to find a number of bugs in the inference machine.

Let's take infection – actual bugs – as an example. After detecting a pathogen, your immune system clicks into gear to produce inflammation. This is rapidly communicated to the brain, where it represents surprise or uncertainty: there is a mismatch between the brain's predicted model of the body's immune state and the actual situation on the ground. In order to minimize this error, the brain acts – creating sickness behaviour, releasing hormones and modulating the immune response via nerves. This response should correct the shock to the defence system brought about by the infection. But what if the brain's response does not resolve the situation, and the prediction error remains high?

If the defence system cannot restore its previous balance – perhaps from an unusually strong inflammatory response, or one that lasts longer than expected – your predictive brain begins to believe that you lack control over your body: low 'self-efficacy'. One clear mani-festation of this is fatigue. Persistent fatigue can be viewed as the brain's recognition of its futile battle to restore balance. Or, in other words, the brain's interpretation of how poorly it is monitoring pro-cesses within its body.[22] And as the predicting brain is dealing with persistent, unresolved uncertainty, not even rest relieves this fatigue. An uncertain brain predicts that it needs more fuel, taking energy from the rest of the body. Some experts have referred to this as a 'cerebral energy crisis', which in the long run can lead to all sorts of chronic disease.[23] This loss of self-efficacy – a loss of control – reduces resilience to future stressors, which further destabilizes the defence system.

This process has been explored in much more detail in the phe-nomenon of chronic (or 'persistent') pain. It is well established that terrible, intense pain can persist for years even after an injury has completely healed. The brain, in trying to secure our survival, can be a real pessimist, and it is not uncommon for pain to become 'wired' on the brain, even in the absence of injury. Many believe that this ultimately results from a brain, over time, failing to predict threats appropriately.[24] This is in no way saying that the pain is 'all in your head' – a changing in the brain's wiring is all too real as a neuro-logical process. If anything, this can be empowering – not only knowing that there are real changes in the brain, but also that, with the right support, it can be 'rewired'. One recently devised treatment for this is called 'pain reprocessing therapy', in which patients received nine one-hour sessions of therapy that involved evidence-based edu-cation in how pain can become stuck on the brain, followed by a guided reappraisal of pain sensations, decoupling threat from the pain experience. A 2022 clinical trial of pain reprocessing therapy for people with chronic back pain (most of whom had suffered for roughly a decade) resulted in the elimination of pain in two thirds of the cohort.[25] This pain relief lasted a year after the trial ended. A larger trial in 2023 found that another treatment focusing on thoughts, behaviours and emotions, cognitive functional therapy (CFT), was also effective at treating long-term back pain.[26] The idea that mind and body need to be harnessed together is now starting to bear clinical fruit; it's an immensely exciting time for the field. A targeted intervention on the defence system can be purely 'psychological', yet still biologically potent and, most importantly, highly effective.

The new science of predictive processing and the story of persistent pain might help to inform the understanding of symptoms that linger long after an infection subsides. Research is beginning to support this at a number of different levels, from the biological to the behavioural. There is evidence that short-term inflammation resulting from an infection can mess with the function of NMDA receptors in the brain – the 'doors of perception' that play a crucial role in your brain's ability to update its model of the world based on new sensory information.[27] We saw the extreme result of the predictive machinery being disturbed in the case of Samantha's anti-NMDA encephalitis: vivid hallucinations and fixed delusions. It is possible that inflammation itself could interfere with the brain's updating of predictions about inflammation. My purposefully depressing trip to Cardiff revealed another cognitive result of inflammation and sickness behaviour: focusing on the negative and having less desire to seek rewards. Emerging evidence also suggests that individuals with chronic pain or fatigue have higher levels of prediction error when it comes to detecting signals coming up from the body.[28] Ultimately, all of this could lead to a brain becoming rewired to believe that it is still fighting an infection. This isn't a fleeting idea that we can easily change: it is the brain's unconscious decision about the state of our body. In the same way that many people live with pain from an old injury (most famously seen in phantom limb pain, where real, terrible pain is felt in a body part that no longer exists), perhaps there are many stuck in the awful situation of fighting a phantom infection.

How do we piece this all together? First, we have to accept that there's no silver bullet that will solve everything. We are drawn to pick a side – 'Team Mind' or 'Team Body' – because it simplifies the messiness of biology and the chaos of chronic illness. But to understand the world, and our relationship with it, we need to be able to handle the chaos. The 'defence system' is a tool to approach these problems. It is not a grand unifying theory, but a helpful structure, a guiding principle to understand the complex circular interaction of mind and body. Your defence system is a tripartite union of mind, body (immune system) and microbiome. While at many points throughout this book we have been picturing this system as a loop – to help understand this circular causality – a complementary geometrical metaphor is that of a triangle. Each point on this triangle has a bidirectional relationship with the other two. A shock to one element of this triumvirate reverberates around the whole system. An adaptive, healthy defence system produces a concerted, contained reaction to a threat. However, for one reason or a thousand, a defence system can become unbalanced, each part inflaming the other in an escalating chain reaction. The result is a fiasco of misdirected energy and accumulated damage.

What does the defence system show us about enigmatic, contested illnesses? It tells us that we are not to blame; that causes and effects are always, obviously real, whether we measure them more as 'psychological' or 'physical'. What does the defence system show us about treatment and recovery? It empowers us with the knowledge and hope for change. There is no need to wait for the experts to come up with a magic pill. The only harm in seeking targeted, expert-led psychological support is that it might receive the judgement of those on one side of the West's mind-body divide. But recalibrating a broken defence system should certainly not rely on psychological therapy alone – in the final chapter we will see that balance requires support from different directions.

Finally, what does the defence system say about yourself? Every experience you have is the fruit of your brain's model of the world when presented with the ever-changing reality of your body and the world. To quote the prince of predictive processing, Professor Karl Friston, we are 'causally embedded in a world through an embodiment of our brain and its cognition'. You are a trinity of brain, body and environment; of mind, immune system and microbe. Instead of having to choose between viewing yourself as a mind or a body, see yourself as a community. That's what a human is.

We have almost reached the final part of the book, in which we will explore the many ways in which we can go about building resilience and restoring damaged defence systems. But, before that, we need to face up to the final, irreversible consequences of a defence system gone rogue.

10

The Cost of War

Chronic inflammation and its fruit

> *This is the way the world ends*
> *This is the way the world ends*
> *This is the way the world ends*
> *Not with a bang but a whimper.*
>
> T. S. ELIOT,
> *THE HOLLOW MEN*

'Tim! How lovely to see you!'

'I'm not Tim,' I replied gently. 'I'm Dr Lyman. I've come to see how you're doing this morning.' The octogenarian ex-cabbie looked sceptical, undecided as to whether I was his son or not.

'A doctor? What's wrong with me?'

'Well, Dennis ... where to begin?' With a mischievous grin, I dramatically flourished a thick wodge of medical notes from the end of his hospital bed. He burst out laughing.

Dennis had been on the geriatrics ward for over a month, but he would meet me for the first time every day. I can't count the number of times I performed similar routines in my junior doctor years, which included sixteen months on geriatric and old-age psychiatry wards. As a doctor, on some level you have to become distanced from the tragedy of dementia, otherwise it would be overwhelming. For that very reason, I've chosen an example from one of my patients, instead of one of my relatives who also suffered with the condition. Dementia affects us all, whether we will experience it ourselves or care for those with it. After cancer, it is the condition we most fear.[1] Unlike cancer, we are yet to see any breakthroughs in its treatment, but recent research into this awful disease – from the population level to the individual, down to the microscopic – is beginning to reveal the fingerprints of inflammation. We are starting to see that, over time, an unbalanced defence system can lead to chronic inflammation, and ultimately the destruction of the structure and functions of the brain, otherwise known as neurodegeneration.

The idea that the immune system could play a role in dementia lay at the fringes of medical science for the best part of a century. One clue, however, was 'delirium'. This is a short-term, reversible state of confusion, which anyone who has worked in healthcare, particularly with the elderly, has seen countless times. Delirium can be triggered by a wide range of things – from sleep deprivation to medications – but one of its most common causes is the inflammatory response that accompanies infection. From my experience in hospitals, pneumonia and urinary tract infections are the usual culprits. Delirium is not dementia, but clinicians have long been aware of the link between the two. Those with dementia are much more vulnerable to developing delirium than the general population. This makes sense: a brain that is starting to move along the miserable path of neurodegeneration is less likely to be resilient when faced with acute stressors. But, curiously, delirium seems to be a strong risk factor for developing dementia in the first place.[2] A 2017 study argued that delirium was a potential cause of dementia, rather than just a correlation. The authors compared the state of almost one thousand autopsied brains with their medical records, finding that delirium accelerates the cognitive dysfunction of dementia well beyond the expected trajectory of decline.[3] What could be accelerating dementia in these cases? Or – the trillion-dollar question – what causes dementia in the first place?

Study by study, it is becoming clearer that inflammation likely plays a significant role. It certainly isn't the only cause of every case of dementia; if there is anything we have learned from the past few chapters it is that complex neuropsychiatric conditions emerge out of a selection of ingredients from our environment and our genetic makeup. But it is now evident that chronic inflammation may well be at the root of dementia.

Let's start by looking at by far the most common and most studied of all dementias: Alzheimer's disease. This is what Dennis had. The hallmark of Alzheimer's disease is short-term memory loss. I've treated some patients who can recite the name of every soldier who fought alongside them on D-Day, and others who remember the price of a pint of milk in 1961, but cannot for the life of them remember what they had for breakfast, or whether they had any breakfast at all. As the disease progresses, memory worsens and other symptoms begin to appear: apathy, mood swings, disorientation, as well as an inability to plan and perform basic tasks. For an observer it looks like someone is gradually going around rooms of the patient's brain, switching the lights off as they leave.

The disease is named after the German psychiatrist Alois Alzheimer, who wrote the first clear documentation of the disease in 1901, on a fifty-year-old woman named Auguste Deter. Her disease steadily progressed and she died in 1906. The first major breakthrough in our understanding came in the 1980s, when it became clear that something very strange was happening in the Alzheimer's brain. Autopsies revealed thick, messy clumps of protein in the neuron-dense grey matter. These aggregates, which form outside of neurons, are made up of the amyloid beta protein and are known as amyloid plaques. Soon another abnormality in the Alzheimer's brain was identified: 'tau tangles'. These are deposits of abnormally aggregated 'tau' proteins, which usually contribute to the structure of nerves in the brain. The classic hypothesis is that amyloid plaques appear first, which then leads to a build-up of tau protein. Despite decades of research, we still know little about how these protein plaques are formed and how they cause dementia. There is little doubt that amyloid plaques play a role in the disease, and genetics suggest that they may play a causal role. The gene for amyloid protein is found on chromosome 21, and those with an extra copy of this chromosome (those with Down's syndrome) are at a much greater risk of developing early-onset Alzheimer's disease, with many starting to develop symptoms in their forties. Another gene associated with an increased risk of Alzheimer's disease is the APOE4 gene. The protein this gene produces usually breaks up amyloid beta protein, and some of those who have dysfunctional APOE4 lack the ability to break up these plaques in the brain.

The amyloid hypothesis is compelling, and drug companies have spent the past few decades engaged in an arms race to develop a cure for Alzheimer's that targets amyloid plaques. The good news is that there have been a number of innovative drugs developed that are essentially antibodies against amyloid beta and that clear away the plaques. The bad news is that none of these have successfully reversed or slowed down the disease, and 2022 was an *annus horribilis* for those hoping for a cure. The pharmaceutical giant Roche's gantenerumab and Genentech's crenezumab fell at the final hurdle in their large human trials, and it was becoming clear that Biogen's aducanumab – which had controversially been approved by the US Food and Drug Administration (FDA) the year before – was not only ineffective but also came with a high risk of brain swelling or bleeding.[4] After thirty years, and tens of millions of dollars, we now have a build-up of ineffective drugs. This does not mean that the amyloid hypothesis has failed. I hope – I pray – that one of these amyloid-attacking antibodies makes a meaningful improvement in those living with this horrific disease. But perhaps treating amyloid beta as the prime suspect stops us from walking up and down the police line-up to consider other potential culprits. Of these, I want us to look into the files of inflammation.

There is plenty of evidence to suggest a correlation between dementias such as Alzheimer's disease and excessive, long-term inflammation in the body, known as chronic inflammation. A 2010 meta-analysis (an analysis of multiple papers, combining their findings) of 1,500 individuals found that those with Alzheimer's disease tended to have raised levels of inflammatory cytokines in their blood.[5] Curiously, further studies found that levels of systemic inflammation tend to be high in the early stages of the disease but not in advanced dementia.[6] We also know that suffering from multiple infections increases the risk of developing dementia.[7] There is also a dose-response relationship: the more infections (regardless of type), the higher the risk of dementia.[8] An intriguing study, published by researchers at Stanford University in 2023, points the finger at one specific infectious agent: the varicella-zoster virus.[9] This is the form of herpes virus we met in the last chapter, which has the dishonourable role of causing both chickenpox and shingles. The team analysed data from the National Health Service in Wales, because in late 2013 the Welsh Government enacted a health intervention that doubles up as a large natural experiment: they rolled out the shingles vaccine to people born on or after 2 September 1933. Over a seven-year follow-up comparing the vaccinated to the unvaccinated, they found that the shingles vaccine reduced the chance of developing dementia by around 20 per cent. While these are early days – and this study raises as many questions as it answers – it is looking likely that infectious agents are responsible for some proportion of dementia cases. Non-infectious inflammatory stimuli also increase the risk of developing dementia, from surgical operations to chronic autoimmune diseases.[10] A remarkable link between systemic inflammation and dementia was uncovered in 2016, when researchers at the University of Southampton found that those with gum inflammation (periodontitis) had a six-fold increased risk of developing Alzheimer's disease over a six-month period.[11] In summary: it appears that inflammation in the body can drive the development of Alzheimer's disease.

Another 2016 study also uncovered a fascinating link between infection and the classical model of Alzheimer's disease. A Boston-based group found that amyloid beta is not a junk protein: it is a natural antibiotic.[12] Clumps of amyloid beta protein are able to entrap bacteria, as well as having antifungal and antiviral properties. Mounting evidence suggests that the presence of these deposits in Alzheimer's could be the end result of an excessive and prolonged inflammatory response in the brain. In September 2022, Donald Weaver, a neurologist and professor at the University of Toronto, published a provocative but persuasive review arguing that Alzheimer's disease is, ultimately, an autoimmune disease.[13] We know that inflammation in the body – whether from an infectious source or not – can affect the brain. But what is becoming increasingly clear is the fact that the ageing brain tends to produce excessive responses to inflammation in the body. Key players in this are the brain's resident immune cells: microglia. While these play a variety of different roles in the brain – both constructive and destructive, from helping clear amyloid plaques to releasing inflammatory cytokines – it appears that when stimulated by systemic inflammation they can become 'primed', reacting more aggressively and producing more inflammation the next time they are exposed to the same inflammatory stimulus.[14] Remarkably, microglia can stay primed for months, perhaps years, developing a long-lasting memory of systemic inflammation.[15] At some point, in some people, a critical mass is reached at which microglia are responding increasingly destructively to inflammation they have helped create. What started as chronic inflammation in the body has become a runaway train of neuroinflammation, resulting in the destruction of neurons and, ultimately, the dysfunction of brain and mind.

Many other lines of research implicate microglia in the development of Alzheimer's and other neurodegenerative diseases. In Chapter 2 we came across Beth Stevens, the Harvard-based pioneer of microglia biology. In a series of experiments published in 2016 and 2017, her lab discovered that – in the brains of mice, at least – microglia not only play an important role in the early development of Alzheimer's, but they do this in a curiously destructive way. Well before cognitive impairment becomes evident, synapses (the junctions between nerve cells) become coated in complement proteins, components of the immune system that are essentially 'EAT ME' signals to microglia, who then proceed to gorge themselves on brain connections.[16] Interestingly, when mice were bred to lack these complement proteins, they showed no synapse loss or cognitive decline.[17] Another line of research linking microglia to Alzheimer's is genetics: genes that confer a risk for developing dementia include those that encode microglial and immune proteins.[18] One fascinating study, published in 2021, linked inflammation with amyloid beta as potential co-causers of Alzheimer's disease.[19] It used advanced brain-scanning on 130 people across different ages and stages of Alzheimer's disease, and found that brain inflammation and extent of amyloid plaque build-up led to damaged neurons and the symptoms of Alzheimer's disease. Perhaps future treatments should target both inflammation and amyloid plaques. Crucially, the recent paradigm shift in our understanding of Alzheimer's disease is that neuroinflammation is not just a consequence of the brain destruction brought about by these protein plaques and tangles: it is a cause. This provides hope for prevention. Evidence is beginning to show that reducing and preventing chronic inflammation – something that is eminently possible through various lifestyle interventions we will encounter in the next chapter – might go some way towards preventing Alzheimer's disease.

As we've seen throughout this book, you cannot explore chronic inflammation and its effects without considering the gut microbiome.

It is becoming evident that individuals with Alzheimer's disease have gut microbiomes that are less diverse than they should be.[20] Whether dysbiosis in dementia is a cause or consequence of the disease is debated, and we need more research. But there is considerable circumstantial evidence from multiple directions that suggests that an unbalanced gut microbiome can drive chronic inflammation in the body and the brain.[21]

There are other tantalizing glimmers of hope in recent research. In Chapter 5 we came across a remarkable proof-of-concept animal study, in which John Cryan's team at University College Cork reversed age-related cognitive decline in mice by giving them the gut microbiome of a younger mouse (in the form of a faecal transplant).[22] Who knows whether we will be transplanting youthful poo into our elders at some point in the near future, but we can have every reason to believe that living in a way that is good for your gut microbiome should help to reduce your risk of developing dementia.

While Alzheimer's disease accounts for around two thirds of all dementia cases, inflammation might also lie at the root of other forms of neurodegeneration. Some of the most tragic cases of dementia I have dealt with have been individuals affected by frontotemporal dementia. As the name suggests, the disease mainly affects the frontal and temporal lobes of the brain, bringing about a wide range of effects, the most prominent being a dramatic personality change. One of my patients, in his late sixties, had been a pleasant, 'rather unexciting' (to quote his wife) suburban accountant. Over the course of just nine months he had become an aggressive, hypersexual 'thug' (to quote his wife again), and he came to our attention after exposing himself in front of a waitress at their local Italian restaurant. When the manager came over to the table, my patient responded with a head-butt. Another strange tragedy of this behavioural variant of frontotemporal dementia (previously known as Pick's disease) is that memory tends to be spared. Patients know what they have done – though often feeling little remorse – and the disease cruelly gives them the illusion of agency over their new-found personality. Loved ones feel as though they are living with a completely different person. A 2020 study from a team at the University of Cambridge scanned the brains of patients with frontotemporal dementia to look at both the presence of misfolded tau protein (similar to those seen in Alzheimer's disease) and inflammation.[23] Interestingly, levels of inflammation in the brain were tightly associated with the amount of abnormal protein.

Another common type of dementia – in fact, the second most common after Alzheimer's disease, making up around 20 per cent of dementia cases – is vascular dementia. 'Vascular' refers to blood vessels, and this kind of dementia is, essentially, caused by a series of mini-strokes. These episodes of poor oxygenation to areas of the brain result in a more dramatic initial onset of dementia, and the disease then tends to progress in a step-by-step fashion. Risk factors for vascular dementia are the same risk factors for stroke – and risk factors for stroke are mostly risk factors for the narrowing of blood vessels: atherosclerosis. It was long assumed that atherosclerosis was simply the result of an imbalance of fat and cholesterol in the blood, resulting in the deposition of fat on artery walls, furring them up and eventually blocking them. It turns out, however, that the driving force behind atherosclerosis is actually chronic inflammation. In their 2021 review article, world experts on atherosclerosis Peter Libby and Oliver Soehnlein state that the process of atherosclerosis 'involves inflammation from its inception to the emergence of complications'.[24]

I want to make a very important point here. Chronic inflammation – which can be viewed as the result of an imbalanced defence system – lies at the root of physical illness as much as it does mental. In fact, atherosclerotic disease is the primary cause of death and disability in the developed world.[25] And inflammation's effects on the body are not limited to atherosclerotic diseases;

inflammation can cause or contribute to cancer, diabetes, auto-immune disorders, osteoporosis and non-alcoholic fatty liver disease.[26] Most of these conditions are diseases of modernity, fostered by a combination of our increasing life expectancies and prolonged exposure to a pro-inflammatory world. We have also seen on our journey so far that inflammation is the Faustian pact your body has made with your immune system: you have an army that protects you from microbial invaders, but healthy tissue is often caught in the crossfire. This is not usually an issue in the first decades of life – your reproductive years – but after that your body and brain begins to pay its debt. Chronic inflammation is the collateral damage from countless microbial wars. It tends to compound when you age, and it is now apparent that this process – termed inflammageing – lies behind many of the ailments of old age, from dementia to heart disease.[27] If one spends enough time in a world of processed food, inadequate exercise, disrupted sleep, the extinction of healthy gut microbiomes, all of the inflammatory consequences of urban living and constant, slow-drip stress, one has a very high chance of developing an unbalanced defence system. Whether affecting mind, body or both, chronic inflammation lies at the root of most modern maladies.

We'd all like a miracle drug and, while a cure for dementia is not on the immediate horizon, there is reason to hope. Alongside the pipeline of drugs targeting the protein plaques and tangles in the Alzheimer's brain, there are also some drugs in development that target overactive microglia and neuroinflammation.[28] Perhaps one of these, or a combination, may one day stop or even reverse the effects of dementia. There is also the possibility that drugs dampening down peripheral inflammation may help. Medications that target the potent pro-inflammatory cytokine TNF-α have revolutionized the treatment of inflammatory conditions such as rheumatoid arthritis and psoriasis, and curiously they also seem to reduce Alzheimer's risk in people with these conditions.[29] Following promising animal research and pilot studies in humans, trials are underway to see if these anti-TNF medications could treat Alzheimer's disease.[30] Similarly, probiotic drugs targeting the microbiome and faecal transplants may also play a role in the treatment of dementia in the future.

We don't currently have a cure for dementia, but that doesn't mean we're helpless. There are, of course, some things we can't control. We're given our genetic make-up at birth, and none of us can avoid the passage of time. Indeed – something that should be said at least once in a health or medical science book: none of us can avoid death. But it's becoming clear – from large population studies to microscopically detailed neuroscience – that inflammation is a significant upstream cause of neurodegeneration. In many cases it could be the most significant cause. And this brings hope, in the fact that chronic inflammation is, to varying extents, preventable and modifiable.

In the first part of the book we learned how to view our mind, immune system and microbiome as a combined defence system. In this second part we have seen how an unbalanced, overprotective defence system affects mind and body, resulting in psychological and physical distress, as well as chronic inflammation. We are about to discover how to rebalance our defence system and come to live – mentally and physically – healthy lives. To finish with an analogy: being a complex, finely engineered system, your body is like a car. Activating the defence system is like hitting the accelerator to get you out of a dangerous situation. If the engine is constantly being over-revved, however, you know the vehicle will quickly pay the price, even if you don't yet know which part will fail.

It's clear that the defence system is a remarkable biological apparatus, but our modern world finds many ways to conspire against it. It's about time we looked at how to build, or rebuild, a supersystem.

PART THREE

Resetting Your Defence System

11

The Anti-Inflammatory Life

I bend so I don't break.

ANON

In the first part of this book, we uncovered the new science behind how your brain, your immune system and your microbiome form a unified defence system. Each influences the other: damage to one sends ripples across the rest of the system.

This has profound implications for how we understand disease.

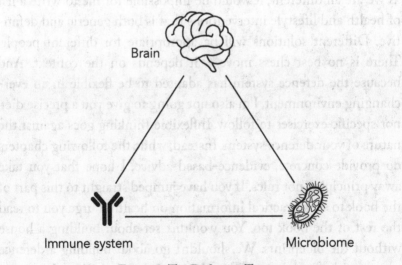

Figure 2: The Defensive Trinity

Body affects mind and mind affects body. Every 'mental health condition' is also physical, and every 'physical health condition' has a mental element. Throughout history, some have already suspected that one might affect the other, but for the first time we are beginning to understand the mechanisms by which they interact. The triumvirate of brain, immune system and microbiome – of mind, body and microbe – is not just interesting; it is a way of moving beyond the polarizing debate that plagues most chronic illnesses. In the second part of the book, we saw how an unbalanced defence system leads to physical and mental distress in the short term, and accumulates collateral damage in the form of chronic inflammation in the long term. This final, practical, part of the book will explore how you can build, restore and maintain a resilient defence system – one that maximizes your wellbeing now and helps prevent disease in the future.

Some caveats

I want this book to excite and empower you. I do not want to tell you what to do. This is neither a self-help book nor a textbook, and, as we are all different, it would be impossible for me to write a list of health and lifestyle interventions that is both generic and definitive. Different solutions will be appropriate for different people. There is no best chess move – it depends on the context. And because the defence system has adapted to be flexible in an ever-changing environment, I'm also not going to give you a precise diet, nor specific exercises to follow. Inflexible thinking goes against the nature of your defence system. Instead, while the following chapters do provide concrete, evidence-based advice, I hope that you take away principles, not rules. If you have jumped straight to this part of the book to seek practical information on health, I urge you to read the rest of the book too. You wouldn't set about building a house without the blueprints. We shouldn't go about building a defence system without understanding why it exists or how it works.

This section is not going to focus on medications. Throughout the book we have had glimpses of exciting new immune-targeted treatments for mental and physical conditions, and the same goes for microbiome-modifying probiotics. But these will need to be tailored for specific conditions and specific individuals. Along with many other clinicians and scientists, I believe there are a number of evidence-based lifestyle interventions at your fingertips. These are interventions that can bring about a balanced defence system – one that begets good health. Much of modern medicine is reactive, but it's becoming clear that the best approach to the diseases of the modern world – heart disease, dementia, diabetes, depression, stroke – is one that is proactive.

While these final chapters are not exhaustive, I hope that they act as a springboard to explore further scientific studies and different ways in which you can achieve meaningful wellbeing – whatever that might mean to you.

On metaphors

As we look at ways in which we can go about building a healthy, resilient mind and body, it helps to keep it all in the context of your defence system. Visualizing this supersystem is not easy, particularly in a society unfamiliar with the concept, so that's where metaphors come to the rescue.

The first metaphor we came across in the book was 'The Ministry of Defence'. If your body were a nation, one of the most important governmental departments would be that which is tasked with the defence of the realm. Some of the civil servants in the Ministry of Defence work in specific areas of the brain, recognizing, reacting to and remembering threats. Others are the infantry of the immune system, patrolling your body's boundaries for intruders. And then there are the foreign delegates, migrants and mercenaries that make up your gut microbiome. While these are not your own cells, they are an invaluable working population that enriches the body and protects it from enemies. To recognize and respond to threats – whether microscopic or macroscopic – this Ministry of Defence must, above all things, be *balanced*. Any defence force needs to tread a tightrope: if the training is undisciplined or an underfunded army is lumbered with cheap equipment, criminal gangs and enemy forces can easily find a foothold. Conversely, if the Ministry of Defence becomes too paranoid or allows its soldiers to be trigger-happy, society descends into a police state – at best, it saps resources and energy from other departments; at worst, it attacks its own citizens. You want a ministry that is well calibrated – neither underactive nor overactive. A defence system should be well trained and ready to fight genuine threats head-on. This gives you security in peacetime, rather than lashing out at everything that passes. So, in this metaphor, many of the ways in which you can strengthen your defence system are essentially about training.

We then came across the metaphor of the bicycle wheel. Your defence system is a spinning wheel, propelling you along the road of life. Part of the wheel is your mind, part is your immune system and part is your microbiome, all linked in a closed circle of causality. As you trundle along the road of life, it is evident that some things are out of your control: the frame of the bike (genetics), and the bumps you have sustained from the road behind you (past infections, experiences and trauma). Over time, these bumps and knocks could slowly tip the wheel out of balance. This creates a friction and heat that threatens to damage the integrity of the wheel and the bicycle as a whole. Having an unbalanced defence system is like trying to ride a bike with a misaligned wheel, and chronic inflammation is the damaging heat created by the wasteful friction. As we've seen throughout the book, chronic inflammation drives poor mental and physical health in the medium term, and in the long term is sustenance for the so-called 'four horsemen of chronic disease': atherosclerotic disease (heart attacks and strokes); metabolic disease (diabetes and obesity); neurodegenerative disease (dementia); and cancer.[1] Facing the road ahead, there will be obstacles that are hard to steer around, but there are some we have the agency to avoid. The rest of this book will look at areas we can potentially change to rebalance the wheel: diet, movement, sleep, interactions with others and, perhaps surprisingly, how you think about yourself.

Take the metaphor that works best for you. Or a bit of both. Or even create your own – anything that helps to remind you that your defence system prioritizes balance. With this balance – as we will see in this final part of the book – comes resilience and adaptability, not perfection. Living an anti-inflammatory life is not about removing inflammation altogether. Like stress, inflammation is a protective weapon. It is instead about learning about how to use this weapon: being confident that it can protect you and being confident that you can hold it in reserve, only using it when necessary. While the science behind the defence system is immeasurably complex, the solutions to building and maintaining a healthy defence system are surprisingly simple. So, without further ado, let's see how you can eat, play and love your way to health.

12

Eat

Variety is the very spice of life.

WILLIAM COWPER,
THE TASK, 1785

When are you most human? Some would say that it is when you demonstrate love and compassion. Others might draw on the adage 'to err is human', arguing that your humanity is most evident when you mess up. I think, however, that you are most human just after you have done a poo.

I'm sure that you were not expecting this revelation in your search for dietary advice, but let's think back to the facts we explored in Chapter 5: 'Your Mind on Microbes'. We saw how your body contains tens of trillions of microbes – as many (and probably more) as you have human cells. Most of these microbes reside in your gastrointestinal tract, forming the gut microbiome. These citizens propagate within your colon, and it's estimated that around half of faecal matter is made up of gut bugs. So it seems perfectly possible that, just for the hours after you've visited the loo, you are more human than microbe.

Your gut microbiome is, essentially, an outsourced organ. With more genetic material than the whole of your body combined, and more metabolic power than your liver, your gut microbiome breaks down the wide menu of foods your own body can't digest – crucial for omnivores like us. Over many millennia, humans have developed a mutual, symbiotic relationship with these microorganisms: we provide a dark, warm and wet habitat, while they confer health and survival benefits to us. One startling detail that underscores the importance of this relationship is the fact that some of the most important nutrients found in breast milk, oligosaccharides, cannot be broken down by the human body. They are instead food designed to shape the infant's nascent microbiome, spurring the growth of beneficial bacteria. In a similar way, a key job of the bacteria residing within your gut is to dine on what I believe to be the most under-rated food group: fibre. While most advice in the dietary world revolves around fats, carbohydrates, proteins and vitamins, few people pay attention to the dull, bulky roughage that is traditionally associated with helping your granny stay regular. Fibre cannot be broken down by human cells, but is instead a feast for bacteria that, as we saw, produce chemicals beneficial to gut, immune system and brain health. Our food is their food.

Perhaps the power of microbes in the food chain can be best demonstrated by a mammal that has mastered fermentation: the cow. I grew up in an English village that held a deep pride in its herd of beautiful chestnut-brown Hereford cattle. As I cycled along the lanes beside their field, watching the brawny beasts ponderously chewing the cud, I would wonder at how a diet of grass resulted in a tonne of bulging muscle. It turns out that the fibrous plants are food for the trillions of bacteria found in the bovine gastrointestinal tract, which in turn produce amino acids, the building blocks of protein that invariably ends up on a plate in the village pub. A cow is essentially a cute, flatulent fermentation vat; a barrel of brewing protein.

I hope that this book has hammered home the message that mind, immune system and microbiome work in concert. The latter is, by far, the easiest to change. There is also ample evidence that building a healthy microbiome leads to a healthy immune system and lays the foundation for a healthy mind. What's not to love? The most powerful predictor of microbiome composition and health is, unsurprisingly, diet – so, this is the perfect place to start.[1] If we want to eat our way to a balanced defence system, we need to first find out what constitutes a healthy microbiome, and then work out how to give it what it wants.

While we don't know the makeup of the 'perfect' microbiome, the data is pointing to a key principle. The microbiomes of healthy individuals tend to have a common denominator: a diversity of microbial species.[2] This is something we see in larger ecosystems across the world. The Amazon rainforest is dripping with life – from the bugs of the soil to the birds of the tree canopy, its variety contributes to the health of the whole environment. Deforestation not only wipes out a multiplicity of species that relied on each other for survival and propagation, its negative effects reverberate across the globe. In the same way, a varied microbiome is more likely to harbour beneficial microbes that provide the body with nutrients and train our immune cells. Perhaps more prosaically, harmless bacteria take up space that might otherwise be dominated by dangerous, opportunistic ones. All in all, diversity begets health.

Natural prebiotics: plant fibre

How to get a diverse microbiome? The answer is astonishingly simple and refreshingly logical: we should eat a diverse diet.[3] In particular, a microbe-friendly diet is one containing a high diversity of plants. A variety of plant fibre encourages the growth of a variety of microbes, which results in the production of a variety of molecules that dampen down inflammation, provide energy to cells and even alter brain chemistry. Your gut microbiome is a natural pharmacy. Perhaps it should be viewed as a medicinal garden.

A short walk from my home, hugged by the sluggish, silty River

Cherwell, lies the University of Oxford Botanic Garden. Britain's oldest physic garden saw the first planting of herbs and flowers for medicinal research, in 1621. Even today, most medicines derive from molecules found in plants, emerging from centuries of trial and error. In a similar way, each species of microbe in your gut is like a medicinal plant that confers specific health benefits. Diverse plant fibre can be seen as the healthy soil that enables these medicinal herbs to thrive. Plant fibre is a natural prebiotic: food for beneficial bacteria.

The benefits of a diet anchored in an abundance of plants is clear to see. The healthiest and longest-lived societies are scattered across the globe, whether it be the mountain villages of Sardinia, the forests of Costa Rica or the Japanese island of Okinawa, but they have one thing in common: a diversity of plants in their diet. The Hadza people of Tanzania are among the last hunter-gatherers on Earth, and they probably consume a diet closest to that which humans have adapted to eat. They forage wild berries, honey and fibre-rich tubers, and eat lean, wild meat. This consumption of between 100g and 150g of varied fibre a day results in a beautifully diverse and robust gut microbiome.[4] This could not be more different from the industrialized West. Americans, on average, eat around 15g of fibre a day – half of the recommended amount and ten times less than the Hadza – resulting in poor microbial diversity.[5] The so-called Standard American Diet (with the apt acronym SAD) has replaced fibre-filled plants with refined grains (plants stripped of their fibre), processed meats, sugar-sweetened drinks and deep-fried food. We all know that the modern Western diet is not good for physical and mental health, but the reason is that it's essentially an anti-biotic diet. Without providing the food for healthy microbes, we end up overfed yet undernourished. Add to this mix an overuse of pharmaceutical antibiotics and a lack of exposure to a variety of environmental microbes due to home-cleaning products and urban living, and no wonder the industrialized world is a desert for microbes. A diet low in plant-based fibre results in a vulnerable microbiome.[6] This increases the likelihood that the defence system loses balance, resulting in chronic inflammation and, eventually, a host of physical and mental health conditions.[7]

If it's clear that a variety of plant fibre is the key to a microbiome-friendly diet, how do we practically implement this? The American Gut Project is a large citizen science project in which individuals across the world volunteer to send stool samples for analysis by a team at University of California San Diego School of Medicine. In 2018, they published the results of over 10,000 participants, finding that eating thirty different plants a week was associated with increased microbiome diversity.[8] This was regardless of whether you were vegetarian or vegan. Most of us in the West manage only around ten plants a week. While not a necessary requirement for good health, thirty a week is a sensible, evidence-based target. It's important to remember that plant foods aren't restricted to fruit and vegetables; nuts, seeds, herbs, spices, grains and legumes also count. While you can add as many layers of complexity as you wish, getting a diverse, healthy microbiome is really as simple as aiming for thirty types of plant a week. One key caveat, however, is to aim to eat plant-based fibre as whole foods, and not in refined, ultra-processed forms. For example, fruit juice has all the sugar of fruit, but with most of the fibre stripped away. In the quest for palatability, snack bars and ready meals often replace plant fibre with nutrient-sparse, but calorie-dense, additives.

One of the most exciting research teams exploring the relationship between gut microbes and their hosts is the Sonnenburg Lab at Stanford University, led by Justin and Erica Sonnenburg. In 2021, this power couple of the microbiome field (they're married) and their team of scientists produced one of the most interesting studies on diet and microbes to date.[9] Over a ten-week period, healthy volunteers were given a diet high in plant fibre, and a detailed analysis of both their microbiomes and immune systems was carried out.

Interestingly, while there appeared to be an increase in the number of healthy microbial products (such as short-chain fatty acids), there wasn't a significant increase in gut microbial diversity across the cohort. This doesn't, however, contradict the evidence that fibre is beneficial for gut microbes and human health. What the stool samples of the subjects showed was evidence of incomplete fibre degradation by microbes. This confirms something that researchers in the USA had previously noticed: people who live in highly industrialized societies and eat Western diets have very low microbiome diversity. They simply don't have the quantity or quality of bacteria to mine the gold that lies within plant fibre. Practically, this might mean that rather than dramatically increasing your dietary fibre intake overnight, you should aim for a slow but steady increase of dietary fibre over a longer time period, allowing for fibre-degrading bacteria to gradually establish a home in your gut. For many of us in the West, this is gut rehabilitation.

I have first-hand experience of this. Throughout my twenties, I had a fairly atrocious diet. This was probably down to two main reasons. The first is arrogance: in my teens I was a committed athlete, competing in sports that required monumental levels of cardiovascular fitness: triathlon, rowing and water polo. I could motor through 3,000-calorie meals without seeming to add an ounce of body fat. I felt that I didn't need to worry about food – something confirmed by the complete lack of teaching on nutrition at medical school. The second reason is that, until fairly recently, I was a terrible cook. One university flatmate recalls me trying to boil pasta without water. While I'm fairly sure that this is an exaggeration, he is not too far off the mark. Around the time I turned thirty and was beginning to explore the science behind the gut microbiome, I decided to revolutionize my diet. In a desperate attempt to bring back the microbes I had for so long neglected, I drastically increased my fibre intake. An immediate attempt to eat thirty types of plant foods in a week just resulted in an irritable bowel and an even more irritable mood. Deducing that I'd slightly overdone things, I decided to start low and go slow. A clove of garlic here, half an onion there. Over the following weeks and months I gradually increased the diversity of plant food in my diet, slowing down if I felt that it was a bit much for my microbes to take. Within a few months I was easily eating around thirty a week, and now I no longer bother counting, as experimenting with different plant foods has become second nature. While I appreciate that I only have a study population of one, this change in diet has had dramatic effects on all aspects of my defence system. I've been plagued with eczema since my late teens – a disease caused in part by immune system dysfunction. Since altering my diet, it has never been better. I have also noticed positive changes in my mood and motivation, as well as a markedly improved resilience to stressors. As with all of us, life happens, and a number of difficult personal events hit me around six months after I'd established a diverse plant diet. These challenges, while difficult, did not elicit the stress responses I mounted to similar struggles a few years previously. A healthy diet is not a magic cure for life's struggles, but it is a fantastic foundation for a resilient defence system.

Before I continue to extrapolate too much from my own experiences, I want to reiterate a key moral of the story: start low and go slow. After that miserable first week, I could have just given up. It makes sense that a gut starved of diversity needs time to build a healthy community. Growing a diverse, healthy microbiome is like working out a muscle: you don't sign up to the gym, then turn up the following day to lift a 100-lb weight and expect to look like Arnold Schwarzenegger. Watching *La Dolce Vita* once does not enable you to converse with the locals on your holiday to Rome (sadly). Nurturing a diverse microbiome requires gentleness, consistency and time. If difficulties arise, a food diary could help. If the introduction of one type of plant food seems to disagree with you, take it back out of your diet, observe what happens following restriction and consider gradually reintroducing it. As with any dietary restriction, this should ideally be supported by a health professional, who would be in a position to diagnose other causes of any discomfort.

While research is continuing to show that a plant-diverse diet is good for your defence system – such as reducing inflammatory cells in the blood via anti-inflammatory changes in the gut microbiome[10] – it's not all down to fibre. Eating a variety of plants is a good way of introducing a wide array of health-conferring chemicals into your body, including vitamins and phytochemicals – molecules that are both food for gut microbes and have some evidence for direct anti-inflammatory properties. Colourful plants tend to contain high levels of these chemicals, so, one way of encouraging this diversity is by trying to 'eat the rainbow'.

A 2023 study carried out by a team at Harvard analysed the findings of eight different diets across a population of over 200,000 people during a thirty-year period.[11] It can be hard to establish clear links between diet and health, but the sheer number of participants in this study certainly helps to establish whether a relationship is real or not. The team found that diets that reduced inflammation resulted in the greatest reduction in chronic diseases such as heart disease, cancer and diabetes. When it compared types of food across the various diets, some foods were wholly bad (such as sugar-sweetened beverages) and some seemed to have mixed results (such as dairy products). A clear finding, however, was that plants and plant-based fibre were overwhelmingly healthy.

While I certainly don't believe that eating a variety of plants is all there is to diet and nutrition, it is a very good place to start. Supporting the invisible ecosystem within your gut also supports the health of our planet, which is already bearing the brunt of deforestation, pollution and climate change. This has led experts across the world to collaborate on the 'planetary health diet', which allows a small amount of meat and dairy but is mostly 'plant-forward': vegetables, legumes, fruits, whole grains and nuts.[12] It might seem somewhat ironic that the focusing on microscopic changes within us might lead to the largest effects for the health of our planet, but we always need to remember that we are not islands: we are part of the chain of life and – as individuals and governments – we have the ability to profoundly alter it.

Natural probiotics: fermented food

The desertification of the Western microbiome and the inability of many to properly break down fibre wasn't the most interesting finding from the 2021 Sonnenburg Lab study, however. The trial was investigating another area: a diet high in fermented foods. These included fermented vegetables (such as sauerkraut and kimchi), yoghurt, kefir and kombucha. Remarkably, a diet rich in these foods improved gut microbiome diversity and dampened down markers of inflammation.

Earlier, we met a man who was a century ahead of his time in studying and speculating about the gut microbiome and immune system: Élie Metchnikoff. In his 1907 book *The Prolongation of Life*, Metchnikoff explains how humans have always been driven to consume foods teeming with microbial life: 'From time immemorial human beings have absorbed lactic microbes by consuming in the uncooked condition substances such as soured milk, kefir, sauerkraut, or salted cucumbers which have undergone lactic fermentation. By these means they have unknowingly lessened the evil consequences of intestinal putrefaction.'[13] He then goes on to explore the use of soured milk in the book of Genesis. The relationship between microbes, humans and diet is a very ancient one. Every culture has a history of fermenting food – from the centrality of kimchi in Korean cuisine to the Icelandic speciality of fermented shark. For much of history, controlled fermentation was one of the only ways of preserving produce. But, as with many things, this rapidly changed in the twentieth century. Preservatives, refrigerators, cans and the pasteurization process have dramatically increased the availability and shelf life of food. But this has come at the cost of halting microbial migration into our bodies. While targeted probiotics – products (such as yoghurts) containing specific microbes – are likely to play a part in microbiome-focused healthcare, perhaps a rediscovery of probiotic foods might help repopulate the desert of the modern microbiome, introducing good bacteria and crowding out the bad ones.

Thankfully, the West is beginning to enter something of a fermentation renaissance. The sourdough craze during the first waves of Covid-19 led to many continuing to make their 'lockdown loaves'. Others, including me, have begun to lovingly nurture their SCOBYs (Symbiotic Cultures of Bacteria and Yeast): jellyfish-like structures that drive the brewing of kombucha. The multicultural range of world foods available in supermarkets and restaurants – miso, tempeh, kimchi, kefir and natto, to name a few – has introduced new life into barren guts. Up to now, this renaissance has been small, middle-class and largely limited to enthusiasts. I decided to speak to someone who is trying to change that. Ironically, it is someone who honed his trade in one of the world's most famous and most exclusive restaurants.

David Zilber's environs could not look more Nordic. His desk is propped up against a wall of glass that frames the ferns, birches and spruces of the Folehaveskoven forest in eastern Denmark. This provides a good counterpoint to the interior of his fermentation lab: sleek, gleaming and chicly clad in exposed concrete. Large glass jars line the shelves, each jar sealed to allow the anaerobic (oxygen-free) growth of microorganisms and bubbling away with gasses released from food being transformed by microbial life. You can see how fermentation got its name: *fervere* being the Latin 'to boil'.

David Zilber is a fermentation scientist at Christian Hansen, a large Danish organization he calls 'a food company with microbiologists instead of chefs'. Initially set up in the late nineteenth century to improve the cheese-making process, Christian Hansen now owns the world's largest collection of bacteria used in the process of making food and beverages. Indeed, Zilber tells me that roughly a third of the calories we eat have been manipulated by microbes in some way.

A chef by training, Zilber's job is now to save the world. He is tasked with the democratization of fermentation, enlivening (literally, as well as figuratively) food products that will find themselves in grocery stores across the globe. He is also using microbes to create delicious plant-based meat alternatives, a pressing need in our fight against climate change. But David Zilber is more than a scientist; he is a culinary artist. His job is not just to explore the health benefits of fermented foods, but to show that they can be utterly delicious. Zilber worked for many years as the Head of Fermentation at Noma, the three-Michelin-starred Copenhagen restaurant that has been ranked 'Best Restaurant in the World' five times by *Restaurant* magazine. I ask him how Noma became known for its experimentation with fermentation.

'It was a bit of a happy accident,' he tells me. 'In the early days of the restaurant, before I joined, the head chef René Redzepi was on the hunt for ancestral ways of foraging and cooking Nordic plants. The team stumbled upon an old Swedish Army manual that helped identify edible plants and ways of storing them. This included a recipe for salting capers to preserve them, which the chefs then decided to do for a range of different berries. After salting some gooseberries and leaving them in a sealed bag at the back of the fridge, the chefs completely forgot about it. About a *year* later, sometime around 2008, one of the chefs happened upon the bag, dipped a spoon into it and brought it to his lips. It was a revelation. It was briny and complex with rich animalistic notes – multiple layers of flavour coming out of what had been an unripe, one-flavour berry.'

The gooseberries had been accidentally lacto-fermented (devoured by healthy *Lactobacillus* bacteria) over the course of a year. This unintentional transformation changed the course of the restaurant.

'The gooseberries were our gateway drug into fermentation,' Zilber explains. 'René set up the Fermentation Lab in 2014, a little bit after I joined as a chef. I would stare longingly at the lab from the heat of the kitchen, wondering what was going on in there. It got to the stage where some of the other chefs thought that I was a bit of a smart-ass and asked if I wanted to work in there.' Self-taught and insatiably curious, Zilber was immediately at home in the Fermentation Lab. 'All I had to do was innovate . . . our only job was to make cool stuff.' Armed with a whiteboard, a diverse supply of ingredients brought in by the foragers and a growing knowledge of differing fermentation techniques, Zilber would act as a shepherd of microbes, directing them to new pastures. 'We started by, say, pickling a specific variety of wild strawberry in every way we could . . . then we'd move on to something else, until we were fermenting everything from lemon myrtles to chapulines (Mexican grasshoppers).'

You don't need to travel to Copenhagen with a reservation at an ultra-exclusive restaurant to enjoy the fruit of Zilber's work, though. He and Redzepi have written *The Noma Guide to Fermentation*: a step-by-step guide to using microorganisms to work as your own sous-chefs. These range from the very simple lacto-fermented plums (essentially salting the fruit and leaving them in a jar for a while) to the more adventurous process of making fermented-grasshopper butter. When it comes to starting out as a home fermenter, Zilber stresses that enjoyment must be the main ingredient: 'Find one gateway drug; one food you enjoy. Whether it's pickling jalapeños for a hot sauce or making a sauerkraut of your favourite vegetables, make sure it's something you love.'

When I ask about the importance of consuming foods rich in microbial life, Zilber is as philosophical as he is scientific: 'One of life's prime directives is to take up space, whether it's one strand of DNA or RNA replicating itself, or one bacteria dividing into two. Any environment is a foothold. Life will creep into any crack. You can't stop life crossing over boundaries, and we couldn't stop microbes occupying our bodies if we tried. So if we can't keep microbes out, we need to host those that we can tolerate, those that benefit us. We need to host microbes that help our body's boundaries – the immune system being a major one – distinguish between what is harmless and what is a genuine threat. In a similar way to how our ancestors domesticated cats to help keep out plague-ridden rats, they implicitly encouraged the growth of bacteria that confer benefits to us. Our culture has been built on the culturing of microbes.'

So, we know that science is beginning to reveal that we have reason to invite these microbes into our lives, and that we need to rewild our bodies. Perhaps that's the reason so many fermented foods taste delicious.

Gut gardening

The Sonnenburg Lab study has catalysed excitement about the possibility that a variety of plant-based fibre and regular fermented food lies at the heart of a healthy, microbe-friendly diet. Researchers are exploring the effects of these diets – alone and in combination – on all aspects of the defence system: microbiome, immune system and mind. When it comes to the latter, John Cryan and his team at University College Cork are calling a combination of prebiotic fibre and fermented food a 'psychobiotic diet', referring to the beneficial effects these microbes have on mental health. In a 2022 pilot study they found that in healthy humans this diet reduces perceived stress.[14] This field is in its infancy, but the future is exciting.

In this chapter I have focused on two dietary changes to improve the health of your defence system: eating a wide variety of plant fibre (aiming for thirty plants a week), and introducing fermented food into your diet. These are not, of course, the only important areas of diet. Exploring the complexities of allergies, intolerances and personalized diets would be a book in itself. Instead, I hope to engender in you a different, positive way of viewing the food you eat. If you accept that half of you is microbial, and understand that you are part of something much greater than yourself, eating becomes truly communal.

I have termed my approach to eating, informed both by psychology and by microbial science, 'gut gardening'. To explain, let's go back to the world of metaphor. If your gut is like a botanical garden full of medicinal life, plant-based fibre is the soil that feeds this life, while eating fermented food is more like planting seeds. Viewing your gut microbiome as a medicinal garden that needs love and attention is a gateway to feeding it, and ultimately your body, the right food. Tending this garden requires gentle care – not everything at once, but a little bit all the time. If you take pride in it gradually becoming more diverse and resilient, not only should you feel more resilient from a psychological point of view, but your brain and body should become healthier through this slow but steady change in diet.

As evidence for the positive effects of microbe-friendly eating for mental health trickles through, perhaps gut gardening could be explored as a form of therapy; a slow introduction of fibre and fermented food consumed in a mindful way. A key tenet of cognitive behavioural therapy (CBT) for depression and disorders of motivation is 'behavioural activation': the very gradual introduction of activities, helping the individual to explore their world and begin to derive motivation from action. It makes complete sense to me to explore different plants in this way, a process that could occur even before the introduction of other activities, such as exercise.

Eating a wide variety of food also has another benefit: a variety of experience. Just as military forces need to train regularly and prepare for different types of warfare, so your immune system and mind need to be exposed to a variety of experiences to stay robust, appropriately reactive and balanced in an ever-changing world. That applies as much to exploring the outside world through socializing, work and exercise as it does to what you put on your plate.

As with almost every area of life, it is probably sensible to follow the 80:20 rule: aim to eat healthily most (80 per cent) of the time, but don't skimp on treats. If you lock yourself into too rigid a diet, or pin your identity on what you eat, you will miss out on the flexibility that your defence system needs. I know that whisky probably confers no health benefits to me, but the occasional tipple to both enjoy the complex flavours and to lubricate conversation with fellow Scotch enthusiasts is absolutely fine. Aim for flexibility and resilience, not perfection. In other words, it's about the general direction of travel, not the destination.

Everything about life and the human body suggests that dynamic stability and growth, not inertia, is the key to health and balance. This is why American author Michael Pollan's evidence-based but non-specific dictum for healthy eating is one of my favourites: 'Eat food. Not too much. Mostly plants.'[15] By 'food', Pollan means real, whole food, not substances processed beyond recognition. We have, after all, adapted to eat food, not nutrients. In a world where eating and diet can be associated with guilt and shame, you should foster a perspective that is kind not just to your microbes, but ultimately to yourself. Sewn into the concept of flexibility is *play*, a concept we will soon dive into. Eating should be pleasurable and, hopefully, sociable. It is a vehicle for exploring your world. As we are about to see, eating may directly drive us – with the help of microbes – to explore.

Key takeaways:

- Feed the garden: eat a variety of plant-based fibre as a natural prebiotic.
  - Diversity is strength! Aim to eat thirty different types of plants a week, which includes fruit, vegetables, herbs, spices, legumes, whole grains and nuts.

- ◦ Eat the rainbow: regularly consume a variety of colourful plants.
- Seed the garden: gradually introduce fermented food into your diet as a natural probiotic.
- Limit ultra-processed foods: these are anti-biotic and, over time, become inflammatory. While a long list is beyond the scope of this book, key culprits include refined carbohydrates (such as white bread), trans fats (high in deep-fried and fast foods), sugary drinks and artificial sweeteners.
- Be mindful of your microbes: you are part of something much greater than yourself: you are a community.
- Be kind to yourself: like exercise or learning a new language, dietary change should be gradual.

13

Play

Now, here, you see, it takes all the running you can do, to stay in the same place.

LEWIS CARROLL, THE RED QUEEN TO ALICE – *THROUGH THE LOOKING-GLASS*

Microbial motivation

It was an early Christmas present. In mid-December 2022, I came across an utterly intriguing scientific paper that presented the discoveries of a research team at the University of Pennsylvania.[1] Their finding: gut microbes motivate mice to move. In their first experiment, they ranked 200 mice by their willingness to exercise on treadmills and running wheels. They then took a series of biological measures from each mouse to find out what influenced motivation to exercise: genes, blood markers, gut microbe composition and metabolic markers. To their surprise, the makeup of the gut microbiome alone could accurately predict exercise motivation. If they removed gut microbes with antibiotics, mice would run less. If they transplanted the gut microbiome of a very athletic mouse into the murine equivalent of a couch potato, the unmotivated mouse would begin to exercise more. It seemed that something in the gut microbiome was stimulating exercise ... but what?

Through a series of clever experiments, they found a pathway.

First, certain gut microbes produce molecules called fatty acid amides (FAAs), which bind to the cannabinoid receptor 1 (CB1) located on specific nerves in the gut. These nerves travel up to the brain, resulting in a surge of the neurotransmitter dopamine during exercise, ultimately increasing motivation for exercise. While we now need to see whether this translates to human studies, it seems that – in mice, at least – bacteria bring about runner's high.

Why this is the case is another matter. Perhaps this detection is a way for the brain to approximate the body's nutritional status, allowing further movement and exercise. Perhaps the bacteria want us to move and explore the world: more exploration, better snacks. More likely, it is the result of a mutual arrangement between man and microbe. Perhaps it is no surprise that the Hadza people of Tanzania – the hunter-gatherers we met in the last chapter, with their wonderfully diverse microbiomes – cover roughly 10 km on foot each day.[2] This is, of course, because they are hunting and gathering, but what they do (and consequently who they are) might also be a product of their microbial makeup. It is well established that movement is tightly linked to eating, but we tend to see this purely through the lens of food consumption and energy expenditure ('calories in, calories out'). Scientists are now beginning to hear the hidden stories of our microbial friends, and how they contribute to our narrative.

You might think it an odd choice to open a chapter on movement and exercise with gut bacteria. But microbiome-friendly dietary changes are the foundation for healthy living on which the rest is built. Reading the University of Pennsylvania paper was such a joy because it offered evidence for my recent personal experience. I spent much of my twenties trying to get back into the running and swimming I had loved so much in my teens. But each year I would give up on my New Year resolutions before January was out. However, a few months after transitioning to a microbe-friendly diet – aiming for thirty different types of plant foods a week and a daily serving of fermented food – I began, almost subconsciously, to enjoy going out for walks. For the first time that I can remember, I began to crave jogging. Every week I would cover a little more ground, aiming to reach the local park, then the old church – each time making it further from my house. Within a few months I had finished my first half-marathon.

More important than my own personal experience is the way that movement is intricately entwined with all aspects of our defence system. Here we have mind, body and microbe in unison, all influencing each other. We tend to see exercise as something we do with our bodies every now and then, a means of improving physical health, losing weight or building muscle, something you take time out of 'real' life to do. But movement is so much more than that. You cannot separate movement from microbes, the immune system, thinking or feeling. It is part of who you are.

Why move?

Knees bend. Lungs expand. Blood pumps. You were made to move. Regular movement, interspersed with rest, is your baseline. The modern world, however, presents the illusion that you can live a life divorced from biology. Whether you are driving to sit in an office all day, or working from home, it is easy for the modern worker to go a whole day having walked only a couple of hundred steps. After spending most of the school day sitting behind a desk, children in Western countries spend about half of their free time sitting down.[3] We have to admit that being sedentary is comfortable, particularly when most of your immediate needs seem to be met. Technological advances mean that we only need to move a thumb to communicate, find entertainment or reply to work emails.

An irresistible body of evidence, however, shows that regular movement is good for us, whereas a sedentary lifestyle wreaks havoc on our physical and mental health.[4] A lack of movement affects everything from mental sharpness to life expectancy. Before we go into any detail, it's important to recognize why movement is necessary for life at its deepest level. Our hunter-gatherer ancestors, whose lives consisted of travelling great distances in search of nutrition, needed complex movement and complex thought to work hand in hand. Foraging and hunting require not just great dexterity and fitness, but great thinking: planning, reasoning, reflecting, imagining and hoping. Thinking is, essentially, internalized movement.

As we've seen, your brain is a prediction machine. All of your experiences are products of both your brain's model – what it expects the world to be like – and sensory feedback trying to report what is actually going on around you. In every moment, the brain is trying to do one thing: reduce uncertainty. This uncertainty relates to the entire world outside the brain, including the body. And the best way of reducing uncertainty in the long run is going out and experiencing the world. Every sensation is a new bit of data that helps refine the brain's model of the world; we use movement to test our environment. A lack of movement can therefore be a causal factor to psychological distress, and a contributing factor to various mental health conditions.

The science is clear that a lack of regular movement can adversely affect your ability to think and reason, as well as your mood. From astronauts experiencing zero gravity to studies exploring sedentary Finnish schoolchildren, it is clear that a lack of movement is linked to low cognitive and academic performance.[5] Movement begets thought, which is probably why our stereotypical image of philosophers has them walking – the Peripatetic school, so named because they moved so much; equally, the Stoics were named after a painted walkway (or *stoa*) in Athens, where they would meet up.

The science is beginning to catch up, revealing many links between movement and cognitive function.[6] The same is true for mood. When – fuelled by my new-found microbial motivation – I began to get into long-distance running, I was amazed by how many people had discovered ultra-running as a treatment for depression. I remember one friend saying, in a tone as philosophical as its content, that 'the process of physically moving forward lets me know that I can move forward with my life'. Indeed, studies have shown that the process of literally moving forward helps people think more about the future and less about the past.[7] The link between thinking, feeling and moving has seeped into our language on many levels: making progress, moving on, moving up.

All in all, the scientific literature is clear about the effects of exercise on mood and mental health. For example, a 2023 meta-analysis examined a huge body of studies looking at various treatments for depression, finding that most forms of exercise, from walking to strength training, were effective.[8] Interestingly, the biggest effect came from dance. The combination of movement, balance, rhythm and social connection is a perfect way to provide the brain with evidence of the outside world. We know much of this intuitively, and this may be a reason why dance is one of humanity's oldest activities. All of this is not to say that talking therapies, medications and diet don't have their place in treating depression. In fact, it might require one or all of these to get someone in a position where they can move forward, literally and figuratively.

Movement is lubricant for the mind, but it is also indispensable for other elements of your defence system. Independent of weight loss, regular movement and exercise are beneficial for almost every aspect of health.[9] One of the key means through which exercise improves health span and life span is by reducing chronic inflammation. It is not just our muscles and joints that are made for movement; our immune cells need us to move regularly too. Sitting or lying still, only interrupted by movement when there is a pressing need, is not the human default. Your baseline should be regular movement interspersed with rest. To frame it another way: being sedentary is inflammatory. Of course I don't mean that the process of sitting down and resting is inflammatory in and of itself. But increased sedentary time is associated with increased inflammatory markers,[10] and replacing just thirty minutes of daily sedentary behaviour with exercise reduces these markers.[11] A thirty-minute walk has also been shown to increase levels of natural killer cells – immune cells that fight viruses and emerging cancer cells.[12]

Among the many studies that show the beneficial effects of movement, I want to focus on a newly discovered mechanism behind how exercise stimulates the immune system. In 2021, researchers in a lab at the University of Texas studied the bone marrow of mice in incredible detail, finding that a specific type of stem cell* gives rise to key immune cells called lymphocytes.[13] But this was not the most interesting finding. The team found that these stem cells contain mechanosensitive receptors, which activate when a mouse is running. This means that exercise is required for the production of these immune cells. To top it all off, they showed that mice that had been genetically altered to lack this system produced fewer immune cells, were less able to clear infectious bacteria and were more likely to die from infection. In sum: movement enhances immunity.

This brings us to a point of confusion that many have with the relationship between exercise and the immune system. We want to have a strong defence system and we don't want chronic inflammation. And we know that exercise activates the immune system. However, we also know that exercise can be inflaming – something everyone who has lifted heavy weights or competed in an endurance event intuitively knows: pain, heat, redness, sometimes even swelling. How do we resolve what looks like a contradiction?

To condense a wide and incomplete literature: in the short term, exercise can activate the immune system and cause inflammation, but in the long term, exercise is anti-inflammatory.[14] Training the

---

* Stem cells called peri-arteriolar LEPR⁺ osteolectin⁺ cells.

immune system – activating it in a healthy way – helps to prevent inflammation becoming chronic. Perhaps it is comparable to the predictive brain, wanting to go out and explore the environment, exposing itself to small amounts of surprise and safe threats so that it can respond better to real threats when they arrive. Exercise and movement essentially trains your defence system.

It is also never too late to start: physically active people in their sixties and seventies tend to have healthier immune systems than inactive people in their twenties.[15] Regular exercise, combined with a healthy diet, reconfigures body tissue: more muscle (which is broadly anti-inflammatory)[16] and less abdominal and visceral fat (which is not, as is commonly assumed, inert, but is without question a pro-inflammatory tissue).[17]

We started our exploration of movement and the defence system with gut microbes, specifically looking at how they might encourage us to move. It is also clear that exercise improves gut microbial health and diversity, regardless of diet. There are probably a number of reasons for this, but it's likely that the strongest is the anti-inflammatory effect of exercise on the immune system in the gut.[18] Just as a healthy microbiome helps balance the immune system, a well-regulated immune system looks after your gut microbes. We are back to our bicycle wheel: our mind, immune system and microbiome are all influencing each other. It can be a vicious cycle, or a virtuous one. This means that we have a number of entry points for healing and strengthening our defence system.

How to move?

Movement is good for the defence system. But how much movement? How often? What type of exercise is best? As with the section on diet, I am not going to be overly prescriptive. We are all different, and life circumstances, disability and culture mean that good movement looks different to each individual. Our understanding of the defence system, however, should lead us to three guiding principles.

The first is that regular movement is good, and some movement is better than none. If you are able to walk or cycle to the shops instead of driving, try to do that as often as is practical. If you can manually sweep and vacuum the floor, don't buy a robot vacuum.

The second is to find and do movement that you enjoy, whether that be an online yoga class, open-water swimming or dancing in the kitchen. If you can incorporate more movement into your life – particularly fun or purposeful movement – do it.

Third is that when it comes to more vigorous exercise, start low and go slow. It's exactly the same approach as a gentle increase of plant diversity in your diet. The goal for most people is one shared by the World Health Organization and many governments across the globe: 150 minutes (two and a half hours) of moderate-intensity aerobic exercise a week. That's the equivalent intensity of a brisk half-hour walk each weekday. Alternatively, the WHO recommends that you could aim for 75 minutes of vigorous intensity exercise – such as running – per week. A sensible goal would really be an equivalent mix of these, spread out over a number of days. The recommendations also advise some kind of muscle-strengthening exercises two days a week.

You can be as creative as you want: movement can be fun and meaningful. For those who are time-poor (myself included) or have sedentary jobs, short 'exercise snacks' are an easy way to fit healthy movement into a busy day. I try to fit in five to ten minutes of press-ups, squats and stretches before my morning shower. To turn these movements into habits, you can use 'triggers' to initiate them; I know someone who does his five-minute morning workout while waiting for his coffee to brew. Or you can 'habit stack', which is simply stacking a new habit on top of one you already do – balancing on one leg while brushing your teeth, for example.

Whenever you have opportunities to take short breaks at work, get away from your desk and aim to move in some kind of way, even if it's a walk around the block. For the average person, the aim is to exercise a little and often. Exercise snacks scattered throughout each day are better than a sedentary week followed by a 5-km run at the weekend. Again, it's the same approach to diet: a healthy salad at the weekend will not offset a week of junk food and sugary drinks. You are what you do.

While the question of how we should move is clearly important, it's easy to neglect the topic of *where* we should move. A 2019 study from the University of Exeter found that spending 120 minutes (two hours) a week in nature is associated with perceived improvements to physical and mental wellbeing.[19] As with the exercise recommendations, this can be broken up into manageable chunks across the week. From 'forest bathing' to breathing in the sea air, we intuitively feel that spending time in nature is good for us, but scientists are only just starting to explore its effects.

Engaging with the sights, sounds and smells of the natural environment is something that brings your focus outwards and connects you to the environment. This playful sensory training of your defence system is likely to improve wellbeing in many ways, and it's no surprise that time in nature is associated with lower stress levels,[20] reduced depression and anxiety,[21] and improved cognitive function.[22] Curiously, it seems that exposure to nature is also good for regulating your immune system.[23] As your defence system is deeply interconnected, part of this is likely to be through reducing stress and improving mental wellbeing. I'm also convinced that exposure to the outside world is good for your microbiome. We are starting to see evidence that time spent outdoors enriches and diversifies your microbiome, and it will be fascinating to see if this is one of the drivers behind the health benefits of green space.[24] In a similar vein, this might play a role in explaining the numerous wellbeing effects of pet ownership. Social interaction, microbe sharing and – if it's a dog – an incentive to exercise in green space is a potent combination.

Move to play

Although we've talked a lot about movement, I believe that the overriding principle the defence system teaches us is *play*. This behaviour is deeply ingrained in mammals and birds, and it permeates all aspects of human culture, from sport to comedy. Play is all about testing your environment. It is about exposing yourself to new challenges that hone your ability to find creative solutions. Play is a conduit to adaptability and resilience. It is, essentially, controlled exposure to a chaotic world. Play can also be viewed as training to face real threats, just as regular 'war games' help put the nation at ease in peacetime.

To physically thrive, you don't need to be an athlete, but you do need to train. Fitness and muscle strength (not necessarily muscle size) is good for mental health as well as physical: it feeds into your brain's perception of what you can achieve and do. At a fundamental level, it is related to your ability to survive when faced with mortal threats. These unconscious messages build up your perception of how capable you are, which is tightly linked to mood. A stronger body leads to more self-efficacy and a feeling that you are in control. In the last few decades there has been an explosion of mental health conditions that, at their core, stem from a loss of perceived control over one's life and future. Perhaps the dramatic increase in sedentary lifestyles in these decades is an under-appreciated factor.

Some notes on rest

There is no single way in which rest can be prescribed to the general population: we are all unique. The pattern of rest for someone training for a marathon is different from that of someone trying to recover from depression, a broken leg or a post-viral illness. The type and amount of rest is context-dependent, but the need for rest is the same. While you should regularly challenge your body and mind, you need time to rest and relax; be gentle and kind to yourself. These things are not mutually exclusive. In the next chapter we will explore practical ways in which you can relax and restore your defence system, resting mind and body. Before then, let's look at a process we think of as rest, but is in fact a wonderfully active and complex physiological process: sleep.

'Routinely sleeping less than six or seven hours a night demolishes your immune system.' This bold claim can be found on the first page of *Why We Sleep*, neuroscientist Matthew Walker's 2017 international bestseller.[25] This book helped millions to wake up, so to speak, to the importance of sleep. A decade or two ago, any mention of sleep in a medical school lecture or doctors' mess would have been met with bewilderment or laughter. These responses unmask the underlying fear that doctors and scientists have of a process – one that effectively involves you spending a third of your life in a coma – that they knew almost nothing about. But now, thanks to pioneering scientists like Walker, we know how and why sleep is fundamental to all aspects of the defence system.

Although it's not an intuitive link, sleep and immunity are deeply entwined. Think back to sickness behaviour: the symptoms one experiences when battling an infection. Fatigue and sleepiness are prominent features of our body's response to infection. Outside of infection, though, sleep is still an immune defence, in which new recruits are conscripted into your immune army and weapon stocks are replenished. In 2015, Dr Aric Prather, one of Walker's colleagues at the University of California, San Francisco, found that people who slept for less than seven hours a night were more susceptible to the common cold virus.[26] Another study found that those who sleep for less than seven hours a night are less likely to mount an appropriate, effective antibody response to flu vaccination.[27] We are also beginning to understand *how* sleep deprivation affects the immune system.

Some key soldiers of the immune army are the natural killer cells, which eliminate both cells infected by viruses and emerging cancer cells. But it appears that even mild sleep deprivation depresses levels of these immune cells.[28] Even one night of four hours' sleep wipes out 70 per cent of natural killer cells in circulation.[29] Sleep loss also appears to increase body-wide chronic inflammation.[30] In short, insufficient sleep leads to the cruel paradox of an unbalanced immune system – one less likely to fight off infections, yet more likely to fight you. The pro-inflammatory state brought about by sleep deprivation is a driving factor behind the long-term results of poor sleep: heart disease, diabetes and dementia, to name a few. To quote Walker again: 'the shorter you sleep, the shorter your life span.'

I've experienced the inflammatory consequences of sleep deprivation many times. By far the greatest trigger for my eczema is a run of hospital night shifts. The effects of night shifts that trouble me the most, however, are those that affect my brain. For almost a whole week following a run of four night shifts in a row, I am morose and irritable, and my cognitive abilities plummet. It is abundantly clear that good sleep is necessary for a healthy mind, from mood to memory.[31] Finally, it also appears that poor sleep is associated with low microbial diversity in the gut.[32] It is likely that the causation is circular (microbes influencing sleep and sleep influencing microbiome composition), so more research is needed to tease out the details. If anything, this should drive us to aim for both good sleep and a healthy gut microbiome.

So, how can we get good sleep? As with diet and movement, there are some low-hanging fruit to pick first. If these are not helping after trying them for four to six weeks, you probably require the more individualized help of a clinician. Getting good sleep starts when you wake up. Try to rise around the same time every morning – as much as you can, of course; I can already hear the laughs of new parents and shift workers. The more regular your sleep rhythm – the consistency of your sleep pattern – the more regulated the chemicals and hormones associated with sleep will be. One of these is adenosine, which gradually builds up throughout the day and makes us feel increasingly sleepy. An unstable sleep rhythm results in alterations in adenosine peaks, which in turn further unbalances sleep rhythm, which worsens sleep ... and on it goes. Try to keep things regular, and resist the weekend lie-in; it'll make you feel less refreshed in the long run.

Another way of supporting your internal body clock – its 'circadian rhythm' – is making use of light. Exposure to bright light, preferably sunlight,* soon after you wake up helps stimulate the natural morning rise in the hormone cortisol.[33] (Perhaps you could use this to 'habit stack', getting your morning sunlight with exercise, in green space if possible.) Although I'm unaware of solid scientific evidence for it, many swear by viewing evening sunlight as well. It makes logical sense, at least, to give your brain and body cues that the day is coming to an end. In a similar vein, try to reduce artificial light around three hours before going to bed, and ideally completely eliminate bright light in the hour before bedtime. When it comes to getting ready for bed, a hot bath can both help you relax and cool your body temperature (due to your body's internal thermostat reacting to the heat by cooling the body), which aids your circadian clock to begin the process of sleep. When it's time for bed, scientists and doctors are clear about 'sleep hygiene', and I particularly like the simplicity of Matthew Walker's maxim: 'dark bedroom, cool bedroom, gadget-free bedroom'.

What you eat and drink also has an effect on how well you sleep. Caffeine temporarily blocks the effects of adenosine, which effectively delays the feeling of sleepiness. As caffeine has a half-life of around five hours (the time it takes for levels to drop by 50 per cent), most experts recommend stopping caffeine intake at least eight hours before you go to bed, ideally longer. Try to resist regular

---

* Although it might seem obvious, please don't look directly at the sun.

alcoholic nightcaps as well: alcohol may help you fall asleep, but there is overwhelming evidence that it ruins sleep quality.[34] Avoid eating large meals too close to bedtime, as this can wake you up in the night, because the process of digestion can trick your brain into thinking that you need to stay awake. Finally, try to go to bed at roughly the same time every night, and aim for around eight hours of sleep. Much of this might seem obvious, but with all the demands on our waking hours it can be surprisingly easy to fall out of the rhythm. Sleep really is medicine and, for most people, these fixes work, if followed. You don't have to do this all at once: start low and go slow. One night of bad sleep is not going to kill you. It's all about the direction of travel.

Play to the rhythm

The ancient-Greek philosopher Heraclitus famously said that 'the only constant in life is change'. Humans have adapted to live in an environment that changes on a daily and seasonal basis. The rhythms of your body are chords written into the symphony of nature. The modern world, on the other hand, presents the illusion that we can rise above the natural order. Ultra-processed food tastes pretty good. If surfing the internet can take us almost anywhere – and deliver almost any item to us – do we really need to move? Now that we can artificially illuminate our lives, what need is there to obey the Sun?

I think that we all have some sense that deviating from biological rhythms is not conducive to good health. Daily and seasonal changes are fairly predictable, but – and this is much more interesting – humans have also adapted to live in an environment that is unpredictable and, frankly, dreadfully chaotic. As your distant ancestor knelt down to drink from their favourite watering hole, there was every chance that the water could have contained cholera. Or that it was now home to a hungry crocodile. Or that it had unexpectedly dried out, resulting in a long walk to find another oasis. To adequately anticipate, recognize and react to threats – whether mammoth-sized or microbe-sized – a human defence system needs to be well stocked and well trained. It needs to be diverse, resilient and adaptable. It needs to be *active*: seeking healthy and diverse nourishment, regularly moving around the environment and actively resting with the rhythms of the day. I stress this activeness to counter the implicit lie we humans have long told ourselves: that happiness and fulfilment is like lying on a cloud, a final state of reclining in comfort and psychological serenity, free of any kind of real effort. My belief is that this is not the case. Instead, as much as it can be achieved, health and happiness is an active state, a dynamic balance. In biology and psychology, the opposite of chaos and stress is not stillness: it is play.

Just to clarify: I am not suggesting that you forgo medicine, forage for all of your food and start practising your cave art. We are privileged to live in the modern world. Enjoy it and make use of modern medicine when it's needed. But, if you want to feel better in the short term and reduce the long-term consequences of inflammation, you need to respect your defence system, the dynamic loop between mind, body and microbe. If we are all mindful of this, and we each make just one positive change to our diet, movement and sleep, society will be transformed.

Key takeaways

- Move regularly. You were made to move.
- Exercise:
  - Aim for 150 minutes of moderate exercise per week, 75 minutes of vigorous exercise, or a combination of the two.
  - A little now is better than none: you can break it down into small 'exercise snacks' throughout the week.

- ◦ Start low and go slow. Be kind to yourself, as each person's circumstances are unique.
- ◦ Play. Make it fun and mix it up. That's an order!
- Get your green time. Where possible, try to spend 120 minutes per week in nature.
- Sleep like your life depends on it:
  - ◦ Alter the variables you can control: timing, light, temperature, and substances like caffeine and alcohol.
  - ◦ If these don't work, seek help from a medical professional.
- Health and happiness are not static goals. They are found in a dynamic balance between brain, body and environment. This is achieved through movement, exploration and play.

14

Love

Hatred paralyses life; love releases it. Hatred confuses life; love harmonizes it. Hatred darkens life; love illuminates it.

MARTIN LUTHER KING,
STRENGTH TO LOVE

I initially thought about naming this final chapter 'Think', as it is primarily about how you can use your mind to help balance your defence system. But I've come to realize that the word 'love' is more appropriate, in ways I hope will become apparent. Here, I mean 'love' in its broadest sense: a love that is kind to yourself and generous towards others. You can't eat, move or sleep your way to good health if you don't have love.

Reach out

I hope that this book has helped you to understand the extent to which mind, body and microbe are intertwined – that your defence system is a loop. So, just as an unbalanced immune system can directly impact mental health, the way you think can have a direct effect on your immune system. This is not magical thinking; it is rapidly becoming established science. Throughout this book – particularly in Chapter 8 – we have seen numerous cases of mind affecting body, from a psychologically stressful situation activating your immune system to new evidence that the brain can 'remember' inflammation and re-inflame bowel tissue with no need for an external, 'physical' trigger. What we term mental health conditions are also deeply physical, and there is no physical health condition that does not have a mental aspect. Psychology is biology. Whether you are recovering from a broken hip or trying to live well with an autoimmune condition, caring for your emotional life is medicine for the body as much as for the mind.

Our society represents a shamefully unhealthy relationship between mind and body: 'physical' is often seen as synonymous with 'real' or 'serious'. By implication, 'mental' and 'emotional' factors are seen as unimportant at best, and imaginary at worst. As we explored in Chapter 1, this thinking is particularly pernicious in the West, due to its enduring reliance on mind-body dualism. It is also a result of an understandable aspect of human nature: a desire for certainty and security. We understandably want our suffering to be validated by measurable scans or blood tests. Doctors also feel more secure in acting on a clear-cut condition; it's comforting to know what solutions to suggest when you're holding an X-ray of a broken bone. We now know, however, that the body and mind are enmeshed in a complex, but beautiful way. That does not mean, though, that healing the mind to heal the body needs to be overly complicated.

The first way to heal through the mind is to ask for help. This can range from seeing a doctor or therapist for a mental health condition, or seeking support from friends, family or charity during life's inevitable lows and seasons of stress. The last thing that I want to do is medicalize normal human emotions, but we should all seek to understand our own emotional landscape and know when to ask for help.

Seeking help does not mean that we are completely helpless. We all have some level of control over our wellbeing, and empowering yourself with more control is the best way to balance your defence system. Part of loving yourself is taking yourself seriously, and understanding your brain and body's extraordinary capability for change. We'll look at how we can tap into that shortly, but for now it is key to know that the more control over your mind and body you can foster, the healthier you will be. This stems from a wide literature that demonstrates the importance of 'self-efficacy' (the belief that you have in your own abilities) as well as having an 'internal locus of control' (believing that you can affect the outcomes in your life). While external forces do play a role in our lives, of course, we all have some degree of agency. Those who believe that they have absolutely no control over their physical and mental health have underestimated themselves and may ironically have given themselves a lower chance of recovery. Of course, there are those who will need much more love and support to regain that sense of control, from victims of childhood trauma to those living in poverty. But, for the time being, let's look at the best ways of using your mind to strengthen your defence system.

Do the basics

There's a reason this chapter comes last. It's certainly not because emotional health is less important than physical health. It's because, for most people, directly working on thoughts and emotions is harder than making changes to diet, exercise and sleep; 'eat more plants' is easier to put into practice than 'learn to manage stress'. More importantly, it is almost impossible to harness the power of your mind when your inputs are a poor diet, a lack of movement and insufficient sleep.

Overall, an overwhelming body of evidence demonstrates that improving the health of your immune system and microbiome is in fact a mental health intervention, as it is all part of the same system. The best way to nurture motivation, mood and mental resilience is to gradually train your defence system through a diverse, plant-led diet, regular playful movement and refreshing sleep. We tend to view resilience as belonging to the mental realm – essentially a reflection of one's character – but psychological resilience is deeply tied to the immune system.[1] And, as we saw in the last chapter, motivation is partially microbial.

Write it down

Once you have started to make changes from the 'physical' end, however, addressing mental and emotional health is not an optional extra. If you are trying to open the lid of a glass jar – particularly if it's stuck – you need to apply a twisting force at both ends. In the same way, you need to mobilize both your body (including its microbial network) and mind. It's easy to fall down a vicious spiral: you don't find the time to exercise, which makes you feel worse and more tired, which reduces your drive to exercise.

The best way of starting to deal with stress and psychological distress is to voice it. To deal with a problem, you need to articulate it first. Voicing it can involve talking with a friend, or it might mean speaking to a clinician. Perhaps the simplest way is to write it down. If you wake up overwhelmed with thoughts of worry about the future, regret for the past or hopelessness in the present, just spend a couple of minutes writing down your thoughts and feelings. There's no right way of doing it: just dump these thoughts on to the page. If you prefer the structure of a journal or diary, use that. Try this daily, at a time of day that suits you best for offloading these thoughts. Over time, this alone can help process difficult feelings and deepen your self-awareness.

If you like, you can add more structure to this process. One example of this is putting your thoughts 'on trial': writing down the situation that has led to the thought, weighing up evidence for both a negative and a positive interpretation of the situation, writing what an independent observer would probably say (or what you would say to a friend in a similar situation), before finally coming to a balanced judgement. All of this is best done with the guidance of a good therapist, but there are free online templates for those with limited access or on long waiting lists.[2]

When it comes to dealing with difficult emotions, putting as accurate a label as possible on each emotion is the best way to address them. A wide literature shows that a better ability to differentiate emotions − not just relying on 'happy' or 'sad' − improves emotional health.[3,4] You can make use of 'emotion wheels', which break down your feeling of 'sadness', for example, into more specific words like 'loneliness', 'despair' or 'abandonment'.[5] Seeking out the right word, like exploring, helps you fine-tune your understanding of yourself.

What you write down shouldn't just be limited to negative thoughts or feelings. One of the most effective 'simple' mental health interventions is 'gratitude practice'. At its simplest, it is voicing − whether writing down, or telling a friend or partner − three things that you are grateful for at the end of each day. You can gradually make this more detailed − the more specific, the better. While the scientific literature on gratitude is young, there is hope that its pro-social effects influence brain and body. A 2021 trial carried out at the University of California Los Angeles found that participants who kept up gratitude practice enjoyed giving support to others, resulting in reduced activity in the amygdala (a key fear centre of the brain) and reduced inflammation in the body. The study's authors came to a conclusion that supports the idea of a defence system learning to become less overactive: 'gratitude may benefit health through the threat-reducing effects of support-giving'.[6] Speaking from personal experience, I have found that saying or writing three things I'm grateful for each day has not only deepened my relationships with others and with myself; it has helped to keep me inspired to reach goals that require persistent effort.

Do things mindfully

Mindfulness is, in its simplest terms, paying attention to the present. This includes present feelings, bodily sensations and thoughts, and learning to hold these without judgement. There is a ton of evidence to support the positive effects of mindfulness on the brain and mental health.[7] Mindfulness is best explored with the help of an experienced professional, but there are some very simple ways to incorporate it into your life by yourself. One of these is through meditation: setting aside five to ten minutes each day to sit in a comfortable chair, close your eyes and selectively direct your focus. This focus direction can be guided by one of the thousands of free mindfulness meditations (or, in one of its recent rebrands, non-sleep deep rest, or NSDR) on YouTube. Find someone who guides your attention to different parts of your body, and encourages deep, slow breathing. The inherent usefulness of meditation, deep reflection and mindfulness is evident from the fact that it is an ancient human tradition, in some form or another, in most cultures. And while you may want to explore this in more complexity, simply finding time to relax and listen to your body is a scientifically sensible thing to do.

Mindfulness can also be accessed in a much more automatic way. Activities that encourage a 'flow state' – a deep, playful absorption in the present – are another way of calming and cultivating a healthy defence system. This can come in whatever form you like: cooking, knitting, DJ-ing or building a model railway. Mindful focus can also be directed towards building other pillars of a healthy defence system. This can range from immersing yourself in the awesome beauty of nature, to contemplating the immense world of microbes you are feeding as you eat. The concept of 'awe' is relatively new to science, but we are beginning to see that wondering at beauty or majesty in the world – and our place in it – is not just beneficial for our emotional health; it is associated with low levels of inflammation.[8]

Finally, mindfulness can be achieved through exercise, not least the Chinese martial art of tai chi. Tai chi fosters relaxation and flow through controlled breathing and body movements. This conveys a wide range of positive health benefits,[9] which is not surprising given that its ancient, enduring philosophy is that defence training should be centred on balancing body and mind. It seems to be a method, honed by trial and error through the mists of time, implicitly designed to mirror and balance your defence system.

From stress to stretch

There is a paradox at the heart of stress. As we saw in Chapter 8, stress is meant to be good: if you bumped into a sabre-toothed tiger on your way to the shops, a brief hyperactivation of your nervous and immune system would be life-saving. Long-term, chronic stress, on the other hand, is a path to chronic inflammation and poor health. Our modern society is, in many ways, one of persistent, chronic stressors, contributing to an already pro-inflammatory environment. We are all exposed to stressful situations, and these are unique to each one of us. It would, therefore, be glib of me to offer a one-size-fits-all approach to dealing with stress. An understanding of the defence system, however, does lead us to establish some principles. The first is to take stress seriously. While you can't – and shouldn't aim to – eliminate stress from your life completely, identifying and removing something that is causing unnecessary, long-term stress is a physical intervention as well as a psychological one. Robert Sapolsky, a Stanford professor of biology and stress expert, argues that, for many of us, the first 20 per cent of our efforts should reduce roughly 80 per cent of stress.[10] Look for low-hanging fruit.

The second principle is to take control of stress. An excessive stress response – a physical and psychological reaction that is out of keeping with the difficulty faced – is often the result of an over-active defence system, one unable to confidently differentiate between safety and threat. Stress researchers argue that there are a number of factors that mediate how we respond to stressful situations. These include having a sense of predictability and control, a perception that life is improving and can improve, and social support. Nurturing a sense of control over your life, building a positive regard for yourself and your ability to change, and seeking love and support from others to help you with this, are absolutely vital.

As we explored in Chapter 11, a police force that is inadequately trained is more likely to overreact and lash out, inflaming and exhausting body and mind. The best way of training a defence system is by going back to basics: gradually introducing a diverse diet, regular movement and adequate sleep. On top of this, employing mental wellness practices will further strengthen your defence system. Learning to do this is just like strengthening a muscle – exposing it to small amounts of safe stress and giving it time to recover and adapt. Bend regularly, so that when troubles inevitably come, you don't break.

The opposite of stress is not tranquillity. It is growth; it is positive challenge; it is play. This is sometimes called 'eustress': positive stress that builds resilience and wellbeing. An example of eustress might involve overcoming a fear of speaking in public, or pushing through nerves to enjoy a first date. Some swear by the eustress of cold showers and ice baths. It is easy to say that we'll go through with these things, but it is hard to go through with them. However, if you live a life mindful of the dynamic, defensive unit of mind, body and microbe that helps you survive and thrive, and if you try to slowly nurture this system, you are on the right path. Stress is not the enemy, but a powerful weapon we need to learn to control. To quote the closing lines of a lecture Robert Sapolsky gave to the University of Illinois: 'To the extent that we are smart enough to have invented psychosocial stressors and then stupid enough to have fallen for them, we all have the potential to be wise enough to keep them in perspective.'[11]

Changing the way we view stress inevitably leads us to the concept of 'mindset'. Your estimation of your capacity for change and growth has an immense impact on your wellbeing and ability to achieve goals. Those with a 'fixed mindset' – believing that human capacities and attributes can't change – are at a huge disadvantage compared with those who can nurture a 'growth mindset'.[12] Obviously, changing the way you think about the world is no easy task. I believe that understanding the defence system is one way that we might change the way we see ourselves, and see how many possibilities we have for growth.

Throughout this book we have witnessed numerous instances of the brain, immune system and microbiome adapting, changing and learning; the defence system places a premium on adapting to an ever-changing environment. There are clearly limitations to the human body – when I was a child, I was frustrated by my inability to grow wings, no matter how hard I tried to wish them into existence. In a more prosaic way, you have no control over the family and society you are born into. That said, each of us has a remarkable capacity for flexibility, a quality that can improve our lives if we take the time and care to nurture it.

When it comes to your physical and mental health, you have to hold on to both realism and hope. Some people are brimming with 'toxic positivity', denying the need to acknowledge pain and suffering. Their message is that we simply need to grin and bear it. In my culture, this appears in the classically British form of the 'stiff upper lip'. The implication is that those who voice their suffering or exhibit psychological distress are weak in character. The result is a culture of blame. At the other end are those who have given in to helplessness. There are many external factors that could have a profoundly negative effect on one's life – abusive parents, an illness or a corrupt government. These factors can justifiably be blamed for the hand you have been dealt in life, but to apportion *all* responsibility of your life to these factors is not the solution – they do not deserve to have the power to determine your future. Ultimately, it is unkind to yourself, as it denies you the chance to meaningfully move forward.

We can all find some level of control and responsibility over our lives, which might require the help of others. To clarify, though, taking responsibility does not equal taking blame. Discerning what we can and can't control is one of the great challenges of being human, and it is deeply rooted in history. We see it in the Eastern practices of mindfulness and tai chi, but it also permeates Western, Judeo-Christian culture. I particularly like the 'Serenity Prayer': 'God, grant me the serenity to accept the things I cannot change, courage to change the things I can, and the wisdom to know the difference.'

A compelling middle way between toxic positivity and fatalistic helplessness is 'tragic optimism'. First defined by the Austrian psychiatrist and Holocaust survivor Viktor Frankl, tragic optimism combines the acceptance of a broken world with hope that despite – and even through – suffering, we can grow.

Get in the habit

There is no panacea for building a balanced defence system. The closest we have are medications that might be necessary for those with severe immunological or psychiatric conditions. Instead, a balanced defence system – and the wellbeing that flows from it – needs to be 'cultivated'. I've chosen this word to reflect a gentle, caring process of growth. To eat, play and love your way to health, you need to act consistently and repeatedly. The Greek philosopher Aristotle noted the importance of habits 2,500 years ago, and his work on the subject was neatly summarized by the American historian Will Durant in 1961: 'You are what you repeatedly do. Excellence, then, is not an act but a habit.'[13]

Thankfully, a recent explosion in the study of habits has provided helpful, evidence-based frameworks for making habits easy and helping them stick. One of the simplest and most accessible popular books on the subject is James Clear's *Atomic Habits*.[14] He argues that to form lasting habits, they need to meet four criteria: obvious, attractive, easy and satisfying.

Let's say that you want to get into the habit of doing some exercise while you wait for your morning coffee to brew. To make it obvious, put a cue near your kettle. It can be as simple as a sticky note on the wall saying 'home gym', or something quirkier, like a little toy model of the Hulk placed on the counter. Whatever you do, make it memorable and fun. Hopefully by doing this – alongside pairing it with pleasurable coffee – you have made it attractive. To make it easy, remember to start low and go slow. It doesn't matter if you do only one press-up or squat on your first day – you have plenty of time to increase this. Finally, make it satisfying both by rewarding yourself with the coffee and by knowing that you are making yourself stronger and healthier.

A 'habit tracker' – whether it be a simple calendar or a fancy app – increases accountability and helps deepen the satisfaction of knowing that you are on the right path. This may all sound fairly straightforward, but it can be hard to do in practice, so holding to a structure that makes a habit as easy as possible is an absolute must. Think about goals you want to achieve, and apply them to this framework. And always remember to be kind to yourself: the goal is progress, not perfection.

Love others

One evening, when I was clearly bored, I asked an artificial intelligence (AI) bot: 'What is the most human thing ever?' A response came back from the void: 'The most human thing ever could be argued to be love, empathy and compassion.' Now, I promise that this is the first time I have used this questionable research method in this book, but the answer is revealing. The reams of literature used to train the machine – which amounts to hundreds of trillions of parameters of online human output – have placed *relationship* at the heart of humanity.

When it comes to health and wellbeing, many would argue that positive human connections are a necessity. Relationships aren't addressed by doctors and scientists as often as they should be, almost certainly because they are almost impossible to measure in an objective way, and they are wonderfully, humanly messy. Having said that, a rich scientific literature confirms what we all intuitively know: social connection is overwhelmingly good for mental and physical health.[15] Connecting with and showing compassion for others is a great way of training and building a healthy defence system on every level. It is also, clearly, very helpful for those most in need.

Some of the biggest predictors of an unbalanced defence system – in the form of chronic stress and chronic inflammation – are loneliness and poverty. This is partly due to the inflammatory lifestyle that poverty or isolation imposes on the individual: poor diet, poor sleep and poor movement. But it is also because, at a fundamental level, the defence system views social isolation and perceived low social status as a threat to life. Neuroimaging studies show that the brain reacts to social rejection in the same way it does to a painful stimulus – even when an online player you have never met before stops throwing you a virtual ball during a computer game.[16] Various studies of animal groups and humans show that those with a low social rank (or low socioeconomic status) are more vulnerable to developing stress-related diseases.[17] This makes perfect sense from a survival perspective. On our own, we humans are relatively weak and vulnerable to predation. Our success has come through the herd, which is built on communication and connection. Literal or perceived isolation from the tribe is, to the defence system, a state of existential threat, vulnerability and stress.

While hierarchies of modern human society are immeasurably complex, those with a real or perceived lack of control in their lives are more likely to have an overactive, unbalanced defence system. This leads to inflammation, which lowers mood and damages tissue in the short term, while, in the long term, its collateral damage manifests in most of the diseases of modern life: heart attacks, strokes, diabetes and dementia. All in all, loneliness and poverty are direct drivers of disease. If Western governments addressed loneliness and poverty in the same way they have successfully diminished another inflammatory stimulus – tobacco smoking – we would all benefit immeasurably.

In short, if any government claims to take health seriously, they must take these aspects seriously, too. For those of us who are fortunate enough to live relatively comfortably and with close social connections, there is little better you can do on this earth than reach out to the isolated and the vulnerable. If each of us were serious about loving our neighbour, the health of our society – in every meaning of the word 'health' – would be transformed.

Love widely

You are in the middle, literally. Without a shadow of a doubt, the coolest fact I learned at primary school (and perhaps the only one I have retained) is that the length of the average human is exactly halfway between the smallest and largest objects in our solar system: 10 billion times larger than an atom and 10 billion times smaller than the Sun. It also appears that, in terms of the biosphere – the interconnected supersystem of all life on our planet – you are also at the halfway point: you are a few million times larger than the smallest bacterium, and a few million times smaller than the Earth.

Whether you like it or not, you are deeply connected to both ends of this writhing mass of life. We are all now acutely aware that human activity is throwing this delicate system out of balance. At current rates, we are polluting, logging and combusting our way to extinction. There is a growing belief, particularly among the young, that if we really care about ourselves and other humans, we have to care for and love the planet as a whole. I completely agree. But I believe that if you want to truly love and respect the world, and the humans in it, you need to direct your love to the other end of the scale as well.

Microbes helped make you, and cultivating a healthy relationship with microorganisms will not only improve your own health, but also help prepare you for taking care of the world. Stewardship of the environment starts with stewardship of your own body. Even if you are eating with no human company, you are providing a feast for millions of microbes, and the food you choose dictates who is invited to the banquet. A mindful approach to microorganisms should go hand in hand with a cultivation of a healthy immune system and a healthy mind. All ecosystems thrive on diversity and balance – the same is true for you. Charity starts at home.

I hope that you have enjoyed the science that has built this book – a new science revealing that mind, body and microbe are indivisibly interlinked. I hope that it inspires you to explore this fascinating subject further, too. But, above all, I hope that this book helps to culti-vate an attitude, one that respects the mind when considering physical health and one that respects the body when considering mental health. You also cannot truly appreciate mind or body without respect for your environment – particularly the invisible microorganisms that call your body home. Keeping this defence system nourished, trained and balanced is the best way of achieving a level of health and well-being that allows you to do the things that you love. Yes, the circular causality of the defence system is more complex than simple, linear narratives of cause and effect – favoured by old school medical text-books and social media health influencers alike. But the truth is worth the complexity. You are not just a mind. You are not just a body. You are a human.

References

Author's Note

1 Edelstein, L., 'The Hippocratic Oath: Text, Translation and Interpretation', *Ancient Medicine: Selected Papers of Ludwig Edelstein*, eds. Temkin, R. and Lilian, C., Johns Hopkins Press, 1967, pp.1484–5

Part One: The Open Mind

1 A Tale of Two Systems

1 Molnár, Z., 'On the 400th anniversary of the birth of Thomas Willis', *Brain*, 144(4), 2021, pp.1033–7. It is worth saying that there is some debate as to the exact location of Willis's dissections. Some argue that they took place a little further east down Oxford's High Street. I'm siding with the conventional theory, though, and I'm sure you'll agree that Chiang Mai Kitchen is more romantic than A-Plan Insurance.

2 Scatliff, J. H. and Johnston, S., 'Andreas Vesalius and Thomas Willis: their anatomic brain illustrations and illustrators', *American Journal of Neuroradiology*, 35(1), 2014, pp.19–22

3 Hughes, J. T., *Thomas Willis 1621–1675: His Life and Work*, Royal Society of Medicine, 1991

4 Molnár, Z., 'Thomas Willis (1621–1675), the founder of clinical neuroscience', *Nature Reviews Neuroscience*, 5 (4), 2004, pp.329–35

5 Shakespeare, W., *The Merchant of Venice: Texts and Contexts*, ed. Kaplan, M. L., Palgrave Macmillan, 2002, pp.25–120

6 Hooke, R., *Micrographia: or Some Physiological Descriptions of Minute Bodies Made by Magnifying Glasses with Observations and Inquiries Thereupon*, Royal Society, 1665, at https://ttp.royalsociety.org//ttp/ttp. html?id=a9c4863d-db77-42d1-b294-fe66c85958b3&type=book&_ ga=2.52409450.239654320.1664281483-1220970289.1663237899

7 Hooke, R., *Micrographia*

8 Pepys, S., *The Diary of Samuel Pepys*, ed. Latham, R., Penguin Classics, 1993

9 McClurken, J. ed., 'Contagion: historical views of diseases and epidemics', *Journal of American History*, 101(4), 2015, pp.1357–8

10 Thucydides, *The History of the Peloponnesian War*, BoD – Books on Demand, 2019; Schaaf, H. S. and Zumla, A., eds, *Tuberculosis: A Comprehensive Clinical Reference*, Elsevier Health Sciences, 2009

11 Needham, J., *Science and Civilisation in China, Volume 6: Biology and Biological Technology; Part 6: Medicine*, Cambridge University Press, 2000. It's important to note that variolation may have independently developed in China, India or Africa, or originated in one and travelled to the others. We just don't know.

12 Jenner, E., *An Inquiry into the Causes and Effects of the Variolae Vaccinae, a Disease Discovered in Some of the Western Counties of England, Particularly Gloucestershire, and Known by the Name of the Cow Pox*, 1798

13 Riedel, S., 'Edward Jenner and the history of smallpox and vaccination', *Baylor University Medical Center Proceedings*, 18(1), 2005, pp.21–25

14 Janeway, C. A., 'Approaching the asymptote? Evolution and revolution in immunology', *Cold Spring Harbor Symposia on Quantitative Biology*, 54(1), 1989, pp.1–13

15 Poltorak, A., He, X., Smirnova, I. *et al.*, 'Defective LPS signaling in C3H/HeJ and C57BL/10ScCr mice: mutations in Tlr4 gene', *Science*, 282(5396), 1998, pp.2085–8

16 Pasparakis, M., Haase, I. and Nestle, F. O., 'Mechanisms regulating skin immunity and inflammation', *Nature Reviews Immunology*, 14(5), 2014, pp.289–301

17 Shirai, Y., 'On the transplantation of the rat sarcoma in adult heterogenous animals', *Japan Medical World*, 1, 1921, pp.14–15

18 Bullmore, E., *The Inflamed Mind: A Radical New Approach to Depression*, Picador, 2018. While it's not clear who first compared the blood-brain barrier to the Berlin Wall, it has been brilliantly popularized by Edward Bullmore, Professor of Psychiatry at the University of Cambridge.

2 The Hole in the Wall

1 Moalem, G., Leibowitz-Amit, R., Yoles, E. *et al.*, 'Autoimmune T cells protect neurons from secondary degeneration after central nervous system axotomy', *Nature Medicine*, 5(1), 1999, pp.49–55

2 Beers, D. R., Henkel, J. S., Zhao, W. *et al.*, 'CD4+ T cells support glial neuroprotection, slow disease progression, and modify glial morphology in an animal model of inherited ALS', *Proceedings of the National Academy of Sciences*, 105(40), 2008, pp.15558–63

3 Pavlovic, S., Daniltchenko, M., Tobin, D. J. *et al.*, 'Further exploring the brain–skin connection: stress worsens dermatitis via substance P-dependent neurogenic inflammation in mice', *Journal of Investigative Dermatology*, 128(2), 2008, pp.434–46

4 Ransohoff, R. M. and Engelhardt, B., 'The anatomical and cellular basis of immune surveillance in the central nervous system', *Nature Reviews Immunology*, 12(9), 2012, pp.623–35

5 Aspelund, A., Antila, S., Proulx, S. T. *et al.*, 'A dural lymphatic vascular system that drains brain interstitial fluid and macromolecules', *Journal of Experimental Medicine*, 212(7), 2015, pp.991–9

6 Bower, N. I., Koltowska, K., Pichol-Thievend, C. *et al.*, 'Mural lymphatic endothelial cells regulate meningeal angiogenesis in the zebrafish', *Nature Neuroscience*, 20(6), 2017, pp.774–83

7 Absinta, M., Ha, S. K., Nair, G. *et al.*, 'Human and nonhuman primate meninges harbor lymphatic vessels that can be visualized noninvasively by MRI', *Elife*, 6, 2017, e29738

8 Iliff, J. J., Wang, M., Liao, Y. *et al.*, 'A paravascular pathway facilitates CSF flow through the brain parenchyma and the clearance of interstitial solutes, including amyloid β', *Science Translational Medicine*, 4(147), 2012, p.147ra111

9 Ringstad, G. and Eide, P. K., 'Cerebrospinal fluid tracer efflux to parasagittal dura in humans', *Nature Communications*, 11(1), 2020, pp.1–9; Rustenhoven, J., Tanumihardja, C. and Kipnis, J., 'Cerebrovascular anomalies: perspectives from immunology and cerebrospinal fluid flow', *Circulation Research*, 129(1), 2020, pp.174–94

10 Rustenhoven, J., Drieu, A., Mamuladze, T. *et al.*, 'Functional characterization of the dural sinuses as a neuroimmune interface', *Cell*, 184(4), 2021, pp.1000–1016

11 Herisson, F., Frodermann, V., Courties, G. *et al.*, 'Direct vascular channels connect skull bone marrow and the brain surface enabling myeloid cell migration', *Nature Neuroscience*, 21(9), 2018, pp.1209–17

12 Cugurra, A., Mamuladze, T., Rustenhoven, J. *et al.*, 'Skull and vertebral bone marrow are myeloid cell reservoirs for the meninges and CNS parenchyma', *Science*, 373(6553), 2021, eabf7844

13 Mazzitelli, J. A., Smyth, L. C., Cross, K. A. *et al.*, 'Cerebrospinal fluid regulates skull bone marrow niches via direct access through dural channels', *Nature Neuroscience*, 25(5), 2022, pp.555–60

14 Del Rio-Hortega, P., 'El tercer elemento de los centros nerviosos. I. La microglia en estado normal II. Intervencion de la microglia en los procesos patologicos. HI. Naturaleza probable de la microglia', *Boll. Socieded Esp. Biol.*, 9, 1919, pp.69–120

15 Penfield, W., 'Microglia and the process of phagocytosis in gliomas', *American Journal of Pathology*, 1(1), 1925, p.77

16 Ginhoux, F., Greter, M., Leboeuf, M. *et al.*, 'Fate mapping analysis reveals that adult microglia derive from primitive macrophages', *Science*, 330(6005), 2010, pp.841–5

17 Pasciuto, E., Burton, O. T., Roca. *et al.*, 'Microglia require CD4 T cells to complete the fetal-to-adult transition', *Cell*, 182(3), 2020, pp.625–40

18 Stevens, B., Allen, N. J., Vazquez, L. E. *et al.*, 'The classical complement cascade mediates CNS synapse elimination', *Cell*, 131(6), 2007, pp.1164–78

19 Schafer, D., Lehrman, E., Kautzman, A. *et al.*, 'Microglia sculpt postnatal neural circuits in an activity and complement-dependent manner', *Neuron*, 74(4), 2012, pp.691–705

20 Lehrman, E. K., Wilton, D. K., Litvina, E. Y. *et al.*, 'CD47 protects synapses from excess microglia-mediated pruning during development', *Neuron*, 100(1), 2018, pp.120–34

21 Woo, J. J., Pouget, J. G., Zai, C. C. *et al.*, 'The complement system in schizophrenia: where are we now and what's next?', *Molecular Psychiatry*, 25(1), 2020, pp.114–30

3 The Sick Sense

1 Eisenberger, N. I., Inagaki, T. K., Mashal, N. M. *et al.*, 'Inflammation and social experience: an inflammatory challenge induces feelings of social disconnection in addition to depressed mood', *Brain, Behavior, and Immunity*, 24(4), 2010, pp.558–63

2 Lasselin, J., Treadway, M. T., Lacourt, T. E. *et al.*, 'Lipopolysaccharide alters motivated behavior in a monetary reward task: a randomized trial', *Neuropsychopharmacology*, 42(4), 2017, pp.801–10

3 Moieni, M., Irwin, M. R., Jevtic, I. *et al.*, 'Inflammation impairs social cognitive processing: a randomized controlled trial of endotoxin', *Brain, Behavior, and Immunity*, 48, 2015, pp.132–8

4 Lasselin, J., Lekander, M., Benson, S., *et al.*, 'Sick for science: experimental endotoxemia as a translational tool to develop and test new therapies for inflammation-associated depression', *Molecular Psychiatry*, 26(8), 2021, pp.3672–83

5 De Marco, R., Ronen, I., Branzoli, F. *et al.*, 'Diffusion-weighted MR spectroscopy (DW-MRS) is sensitive to LPS-induced changes in human glial morphometry: a preliminary study', *Brain, Behavior, and Immunity*, 99, 2022, pp.256–65

6 Garg, R. and Qadri, A., 'Hemoglobin transforms anti-inflammatory Salmonella typhi virulence polysaccharide into a TLR-2 agonist', *Journal of Immunology*, 184(11), 2010, pp.5980–7

7 Axelsson, J., Sundelin, T., Olsson, M. J. *et al.*, 'Identification of acutely sick people and facial cues of sickness', *Proceedings of the Royal Society B: Biological Sciences*, 285(1870), 2018, pp.2017–2430

8 Sundelin, T., Karshikoff, B., Axelsson, E. *et al.*, 'Sick man walking: perception of health status from body motion', *Brain, Behavior, and Immunity*, 48, 2015, pp.53–6

9 Lasselin, J., Sundelin, T., Wayne, P. M. *et al.*, 'Biological motion during inflammation in humans', *Brain, Behavior, and Immunity*, 84, 2020, pp.147–53

10 Bomers, M. K., Van Agtmael, M. A., Luik, H. *et al.*, 'Using a dog's superior olfactory sensitivity to identify *Clostridium difficile* in stools and patients: proof of principle study', *BMJ*, 345, 2012, e7396

11 Lekander, M., *The Inflamed Feeling: The Brain's Role in Immune Defence*, Oxford University Press, 2021

12 Willis, T., *The London Practice of Physick, or the Whole Practical Part of Physick*, Classics of Medicine Library, Gryphon Editions 2011

13 Olsson, M. J., Lundström, J. N., Kimball, B. A., *et al.*, 'The scent of disease: human body odor contains an early chemosensory cue of sickness', *Psychological Science*, 25(3), 2014, pp.817–23

14 Regenbogen, C., Axelsson, J., Lasselin, J. *et al.*, 'Behavioral and neural correlates to multisensory detection of sick humans', *Proceedings of the National Academy of Sciences*, 114(24), 2017, pp.6400–5

15 Gordon, A. R., Kimball, B. A., Sorjonen, K. *et al.*, 'Detection of inflammation via volatile cues in human urine', *Chemical Senses*, 43(9), 2018, pp.711–19; Olsson, M. J. *et al.*, 'The scent of disease', op. cit.; Sarolidou, G., Axelsson, J., Kimball, B. A. *et al.*, 'People expressing olfactory and visual cues of disease are less liked', *Philosophical Transactions of the Royal Society B*, 375(1800), 2020, p.20190272

16 Stevenson, R. J., Hodgson, D., Oaten, M. J. *et al.*, 'Disgust elevates core body temperature and up-regulates certain oral immune markers', *Brain, Behavior, and Immunity*, 26(7), 2012, pp.1160–68; Schaller, M., Miller, G. E., Gervais, W. M. *et al.*, 'Mere visual perception of other people's disease symptoms facilitates a more aggressive immune response', *Psychological Science*, 21(5), 2010, pp.649–52

17 Murray, D. R. and Schaller, M., 'Threat (s) and conformity deconstructed: perceived threat of infectious disease and its implications for conformist attitudes and behavior', *European Journal of Social Psychology*, 42(2), 2012, pp.180–8

18 Murray, D. R., Kerry, N. and Gervais, W. M., 'On disease and deontology: multiple tests of the influence of disease threat on moral vigilance', *Social Psychological and Personality Science*, 10(1), 2019, pp.44–52

19 Aarøe, L., Petersen, M. B. and Arceneaux, K., 'The behavioral immune system shapes political intuitions: why and how individual differences in disgust sensitivity underlie opposition to immigration', *American Political Science Review*, 111(2), 2017, pp.277–94

20 Schnall, S., Benton, J. and Harvey, S., 'With a clean conscience: cleanliness reduces the severity of moral judgments', *Psychological Science*, 19(12), 2008, pp.1219–22

21 Kaňková, Š., Takács, L., Krulová, M., *et al.*, 'Disgust sensitivity is negatively associated with immune system activity in early pregnancy: Direct support for the Compensatory Prophylaxis Hypothesis', *Evolution and Human Behavior*, 43(3), 2022, pp.234–41

22 Navarrete, C. D., Fessler, D. M. and Eng, S. J., 'Elevated ethnocentrism in the first trimester of pregnancy', *Evolution and Human Behavior*, 28(1), 2007, pp.60–5

23 Zakrzewska, M., 'Olfaction and prejudice: the role of body odor disgust sensitivity and disease avoidance in understanding social attitudes', PhD thesis, Department of Psychology, Stockholm University, 2022

24 Zakrzewska, M., Olofsson, J. K., Lindholm, T., *et al.*, 'Body odor disgust sensitivity is associated with prejudice towards a fictive group of immigrants', *Physiology & Behavior*, 201, 2019, pp.221–7; Zakrzewska, M. Z., Liuzza, M. T., Lindholm, T. *et al.*, 'An overprotective nose? Implicit bias is positively related to individual differences in body odor disgust sensitivity', *Frontiers in Psychology*, 11, 2020, p.301

25 Kipnis, J., 'Immune system: the "seventh sense"', *Journal of Experimental Medicine*, 215(2), 2018, pp.397–8

4 A Tale of a Supersystem

1 Summerfield, C., Egner, T., Greene, M. *et al.*, 'Predictive codes for forthcoming perception in the frontal cortex', *Science*, 314(5803), 2006, pp.1311–14

2 George, K. and Das, J. M., 'Neuroanatomy, thalamocortical radiations', StatPearls Publishing, 2019

3 Pace-Schott, E. F., Amole, M. C., Aue, T., *et al.*, 'Physiological feelings', *Neuroscience & Biobehavioral Reviews*, 103, 2019, pp.267–304

4 Barrett, L. F., *How Emotions are Made: The Secret Life of the Brain*, Pan Macmillan, 2017

5 Miller, A. H. and Raison, C. L., 'The role of inflammation in depression: from evolutionary imperative to modern treatment target', *Nature Reviews Immunology*, 16(1), 2016, pp.22–34

6 Bierhaus, A., Wolf, J., Andrassy, M., *et al.*, 'A mechanism converting psychosocial stress into mononuclear cell activation', *Proceedings of the National Academy of Sciences*, 100(4), 2003, pp.1920–5

7 Ader, R. and Cohen, N., 'Behaviorally conditioned immunosuppression', *American Psychosomatic Society*, 37(4), 1975, pp.333–40

8 Bhat, A., Parr, T., Ramstead, M. *et al.*, 'Immunoceptive inference: why are psychiatric disorders and immune responses intertwined?', *Biology & Philosophy*, 36(3), 2021, pp.1–24

9 Lynall, M. E., Soskic, B., Hayhurst, J., *et al.*, 'Genetic variants associated with psychiatric disorders are enriched at epigenetically active sites in lymphoid cells', *Nature Communications*, 13(1), 2022, pp.1–15

10 Miller, A. H. and Raison, C. L., 'The role of inflammation in depression: from evolutionary imperative to modern treatment target', *Nature Reviews Immunology*, 16(1), 2016, pp.22–34

11 Klunk, J., Vilgalys, T. P., Demeure, C. E. *et al.*, 'Evolution of immune genes is associated with the Black Death', *Nature*, 611, 2022, pp.312–19

5 Your Mind on Microbes

1 Boillat, M., Hammoudi, P. M., Dogga, S. K., *et al.*, 'Neuroinflammation-associated aspecific manipulation of mouse predator fear by Toxoplasma gondii', *Cell Reports*, 30(2), 2020, pp.320–34

2 Tauber, A. I., 'Metchnikoff and the phagocytosis theory', *Nature Reviews Molecular Cell Biology*, 4(11), 2003, pp.897–901

3 Bouchard, C., *Lectures on Auto-Intoxication in Disease, or Self-poisoning of the Individual*, trans. Thomas Oliver, F. A. Davis Company, 1906

4 Metchnikoff, I. I., *The Prolongation of Life: Optimistic Studies* (1907), G. P. Putnam's Sons, 1910

5 Sudo, N., Chida, Y., Aiba, Y. *et al.*, 'Postnatal microbial colonization programs the hypothalamic–pituitary–adrenal system for stress response in mice', *Journal of Physiology*, 558(1), 2004, pp.263–75

6 O'Mahony, S. M., Marchesi, J. R., Scully, P. *et al.*, 'Early life stress alters behavior, immunity, and microbiota in rats: implications for irritable bowel syndrome and psychiatric illnesses', *Biological Psychiatry*, 65(3), 2009, pp.263–67

7 Clarke, G., Grenham, S., Scully, P. *et al.*, 'The microbiome-gut-brain axis during early life regulates the hippocampal serotonergic system in a sex-dependent manner', *Molecular Psychiatry*, 18(6), 2013, pp.666–73

8 Bravo, J. A., Forsythe, P., Chew, M. V. *et al.*, 'Ingestion of Lactobacillus strain regulates emotional behavior and central GABA receptor expression in a mouse via the vagus nerve', *Proceedings of the National Academy of Sciences*, 108(38), 2011, pp.16050–5

9 Hoban, A. E., Stilling, R. M., Moloney, G. *et al.*, 'The microbiome regulates amygdala-dependent fear recall', *Molecular Psychiatry*, 23(5), 2018, pp.1134–44

10 Luczynski, P., Whelan, S. O., O'Sullivan, C. *et al.*, 'Adult microbiota-deficient mice have distinct dendritic morphological changes: differential effects in the amygdala and hippocampus', *European Journal of Neuroscience*, 44(9), 2016, pp.2654–66

11 Hoban, A .E., Stilling, R. M., Ryan, F. J. *et al.*, 'Regulation of prefrontal cortex myelination by the microbiota', *Translational Psychiatry*, 6(4), 2016, e774

12 Desbonnet, L., Clarke, G., Traplin, A. *et al.*, 'Gut microbiota depletion from early adolescence in mice: implications for brain and behaviour', *Brain, Behavior, and Immunity*, 48, 2015, pp.165–73

13 Bercik, P., Denou, E., Collins, J. *et al.*, 'The intestinal microbiota affect central levels of brain-derived neurotropic factor and behavior in mice', *Gastroenterology*, 141(2), 2011, pp.599–609

14 Kelly, J. R., Borre, Y., O'Brien, C. *et al.*, 'Transferring the blues: depression-associated gut microbiota induces neurobehavioural changes in the rat', *Journal of Psychiatric Research*, 82, 2016, pp.109–18

15 Clarke, G., Stilling, R. M., Kennedy, P. J. *et al.*, 'Minireview: Gut microbiota: the neglected endocrine organ', *Molecular Endocrinology*, 28(8), 2014, pp.1221–38

16 Paun, A. and Danska, J. S., 'Immuno-ecology: how the microbiome regulates tolerance and autoimmunity', *Current Opinion in Immunology*, 37, 2015, pp.34–9

17 Flegr, J., 'Effects of *Toxoplasma* on human behaviour', *Schizophrenia Bulletin*, 33(3), 2007, pp.757–60

18 Trevelline, B. K. and Kohl, K. D., 'The gut microbiome influences host diet selection behavior', *Proceedings of the National Academy of Sciences*, 119(17), 2022, e2117537119

19 Swartz, T. D., Duca, F. A., De Wouters, T. *et al.*, 'Up-regulation of intestinal type 1 taste receptor 3 and sodium glucose luminal transporter-1 expression and increased sucrose intake in mice lacking gut microbiota', *British Journal of Nutrition*, 107(5), 2012, pp.621–30

20 Ezra-Nevo, G., Henriques, S. F. and Ribeiro, C., 'The diet-microbiome tango: how nutrients lead the gut brain axis', *Current Opinion in Neurobiology*, 62, 2020, pp.122–32

21 Bohórquez, D. V., Shahid, R. A., Erdmann, A. *et al.*, 'Neuroepithelial circuit formed by innervation of sensory enteroendocrine cells', *Journal of Clinical Investigation*, 125(2), 2015, pp.782–6

22 Buchanan, K. L., Rupprecht, L. E., Kaelberer, M. M. *et al.*, 'The preference for sugar over sweetener depends on a gut sensor cell', *Nature Neuroscience*, 25(2), 2022, pp.191–200

23 Kaelberer, M. M., Rupprecht, L. E., Liu, W. W. *et al.*, 'Neuropod cells: emerging biology of the gut-brain sensory transduction', *Annual Review of Neuroscience*, 43, 2020, p.337

24 Zhang, W., Lyu, M., Bessman, N. J. *et al.*, 'Gut-innervating nociceptors regulate the intestinal microbiota to promote tissue protection', *Cell*, 185(22), 2022, pp.4170–89; Yang, D., Jacobson, A., Meerschaert, K. A. *et al.*, 'Nociceptor neurons direct goblet cells via a CGRP-RAMP1 axis to drive mucus production and gut barrier protection', *Cell*, 185(22), 2022, pp.4190–205

25 Dalile, B., Van Oudenhove, L., Vervliet, B. *et al.*, 'The role of short-chain fatty acids in microbiota-gut-brain communication', *Nature Reviews Gastroenterology & Hepatology*, 16(8), 2019, pp.461–78

26 Strandwitz, P., 'Neurotransmitter modulation by the gut microbiota', *Brain Research*, 1693 (Part B), 2018, pp.128–33

27 Kennedy, K. M., Gerlach, M. J., Adam, T. *et al.*, 'Fetal meconium does not have a detectable microbiota before birth', *Nature Microbiology*, 6(7), 2021, pp.865–73

28 Stokholm, J., Thorsen, J., Blaser, M. J. *et al.*, 'Delivery mode and gut microbial changes correlate with an increased risk of childhood asthma', *Science Translational Medicine*, 12(569), 2020, eaax9929

29 Shao, Y., Forster, S. C., Tsaliki, E. *et al.*, 'Stunted microbiota and opportunistic pathogen colonization in caesarean-section birth', *Nature*, 574(7776), 2019, pp.117–21

30 Busi, S. B., de Nies, L., Habier, J. *et al.*, 'Persistence of birth mode-dependent effects on gut microbiome composition, immune system stimulation and antimicrobial resistance during the first year of life', *ISME Communications*, 1(1), 2021, pp.1–12

31 Dinan, T. G., Kennedy, P. J., Morais, L. H., Murphy, A., Long-Smith, C. M., Moloney, G. M., Bastiaanssen, T. F., Allen, A. P., Collery, A., Mullins, D. and Cusack, A. M., 'Altered stress responses in adults born by Caesarean section', *Neurobiology of Stress*, 16, 2022, p.100425

32 Stewart, C. J., Ajami, N. J., O'Brien, J. L. *et al.*, 'Temporal development of the gut microbiome in early childhood from the TEDDY study', *Nature*, 562(7728), 2018, pp.583–88

33 Rodríguez, J. M., 'The origin of human milk bacteria: is there a bacterial entero-mammary pathway during late pregnancy and lactation?', *Advances in Nutrition*, 5(6), 2014, pp.779–84

34 Pereyra-Elías, R., Quigley, M. A. and Carson, C., 'To what extent does confounding explain the association between breastfeeding duration and cognitive development up to age 14? Findings from the UK Millennium Cohort Study', *PLOS ONE*, 17(5), 2022, e0267326

35 Radford-Smith, D. E., Probert, F., Burnet, P. W. *et al.*, 'Modifying the maternal microbiota alters the gut–brain metabolome and prevents emotional dysfunction in the adult offspring of obese dams', *Proceedings of the National Academy of Sciences*, 119(9), 2022, e2108581119

36 Hebert, J. C., Radford-Smith, D. E., Probert, F. *et al.*, 'Mom's diet matters: maternal prebiotic intake in mice reduces anxiety and alters brain gene expression and the fecal microbiome in offspring', *Brain, Behavior, and Immunity*, 91, 2021, pp.230–44

37 Radford-Smith, D. E., Anthony, D. C., Benz, F., Grist, J. T., Lyman, M., *et al.*, 'A multivariate blood metabolite algorithm stably predicts risk and resilience to major depressive disorder in the general population', *Ebiomedicine*, 93, 2023

38 Luczynski, P., McVey Neufeld, K.-A., Oriach, C. S. *et al.*, 'Growing up in a bubble: using germ-free animals to assess the influence of the gut

microbiota on brain and behavior', *International Journal of Neuropsychopharmacology*, 19(8), 2016

39 Sharon, G., Cruz, N. J., Kang, D. W. *et al.*, 'Human gut microbiota from autism spectrum disorder promote behavioral symptoms in mice', *Cell*, 177(6), 2019, pp.1600–18

40 Hsiao, E. Y., McBride, S. W., Hsien, S., 'Microbiota modulate behavioral and physiological abnormalities associated with neurodevelopmental disorders', *Cell*, 155(7), 2013, pp.1451–63

41 Wan, Y., Zuo, T., Xu, Z. *et al.*, 'Underdevelopment of the gut microbiota and bacteria species as non-invasive markers of prediction in children with autism spectrum disorder', *Gut*, 71(5), 2022, pp.910–18; Yap, C. X., Henders, A. K., Alvares, G. A. *et al.*, 'Autism-related dietary preferences mediate autism-gut microbiome associations', *Cell*, 184(24), 2021, pp.5916–31

42 Sherwin, E., Bordenstein, S. R., Quinn, J. L., 'Microbiota and the social brain', *Science*, 366(6465), 2019, eaar2016

43 Kort, R., Caspers, M., van de Graaf, A. *et al.*, 'Shaping the oral microbiota through intimate kissing', *Microbiome*, 2(1), 2014, pp.1–8

44 Claesson, M. J., Jeffery, I. B., Conde, S. *et al.*, 'Gut microbiota composition correlates with diet and health in the elderly', *Nature*, 488(7410), 2012, pp.178–84

45 Scott, K. A., Ida, M., Peterson, V. L. *et al.*, 'Revisiting Metchnikoff: age-related alterations in microbiota-gut-brain axis in the mouse', *Brain, Behavior, and Immunity*, 65, 2017, pp.20–32

46 Boehme, M., van de Wouw, M., Bastiaanssen, T. F. *et al.*, 'Mid-life microbiota crises: middle age is associated with pervasive neuroimmune alterations that are reversed by targeting the gut microbiome', *Molecular Psychiatry*, 25(10), 2020, pp.2567–83

47 Boehme, M., Guzzetta, K. E., Bastiaanssen, T. F. *et al.*, 'Microbiota from young mice counteracts selective age-associated behavioral deficits', *Nature Aging*, 1(8), 2021, pp.666–76

48 Monroy, M. and Keltner, D., 'Awe as a pathway to mental and physical health', *Perspectives on Psychological Science*, 18(2), 2023, pp.309–20

Part Two: Things Fall Apart

6 Friendly Fire

1 Vitaliani, R., Mason, W., Ances, B. *et al.*, 'Paraneoplastic encephalitis, psychiatric symptoms, and hypoventilation in ovarian teratoma', *Annals of Neurology*, 58(4), 2005, pp.594–604

2 Dalmau, J., Gleichman, A. J., Hughes, E. G. *et al.*, 'Anti-NMDA-receptor encephalitis: case series and analysis of the effects of antibodies', *Lancet Neurology*, 7(12), 2008, pp.1091–8

3 'Susannah Cahalan on anti-NMDA encephalitis and her journey to diagnosis', *Brain & Life* podcast hosted by Dr Daniel Correa and Dr Audrey Nath, 28 April 2022

4 Cahalan, S., *Brain on Fire: My Month of Madness*, Simon & Schuster, 2012

5 Buckley, C., Oger, J., Clover, L. *et al.*, 'Potassium channel antibodies in two patients with reversible limbic encephalitis', *Annals of Neurology: Official Journal of the American Neurological Association and the Child Neurology Society*, 50(1), 2001, pp.73–8

6 Zandi, M. S., Irani, S. R., Lang, B., *et al.*, 'Disease-relevant autoantibodies in first episode schizophrenia', *Journal of Neurology*, 258(4), 2011, pp.686–8

7 Pollak, T. A., Lennox, B. R., Müller, S. *et al.*, 'Autoimmune psychosis: an international consensus on an approach to the diagnosis and management of psychosis of suspected autoimmune origin', *Lancet Psychiatry*, 7(1), 2020, pp.93–108

8 Lennox, B. R., Palmer-Cooper, E. C., Pollak, T. *et al.*, 'Prevalence and clinical characteristics of serum neuronal cell surface antibodies in first-episode psychosis: a case-control study', *Lancet Psychiatry*, 4(1), 2017, pp.42–8

9 Scott, J. G., Gillis, D., Ryan, A. E. *et al.*, 'The prevalence and treatment outcomes of antineuronal antibody-positive patients admitted with first episode of psychosis', *BJPsych Open*, 4(2), 2018, pp.69–74

10 Kraepelin, E. (1919), 'Dementia praecox and paraphrenia', trans. E. S. Barclay, in Henderson, D. and Gillespie, R. D., *A Textbook of Psychiatry* (Eighth Edition), Livingstone

11 Lawson, D. (1692), 'A Brief and True Narrative of Some Remarkable Passages Relating to Sundry Persons Afflicted by Witchcraft, at Salem

Village: Which Happened from the Nineteenth of March, to the Fifth of April', *Early English Books Online Text Creation Partnership*, 2011

12 Tam, J. and Zandi, M. S., 'The witchcraft of encephalitis in Salem', *Journal of Neurology*, 264(7), 2017, pp.1529–31

13 Sebire, G., 'In search of lost time from "Demonic Possession" to anti–N-methyl-D-aspartate receptor encephalitis', *Annals of Neurology*, 67(1), 2010, pp.141–2

14 Endres, D., Lüngen, E., Hasan, A. *et al.*, 'Clinical manifestations and immunomodulatory treatment experiences in psychiatric patients with suspected autoimmune encephalitis: a case series of 91 patients from Germany', *Molecular Psychiatry*, 27(3), 2022, pp.1479–89

15 Lennox, B., Yeeles, K., Jones, P. B. *et al.*, 'Intravenous immunoglobulin and rituximab versus placebo treatment of antibody-associated psychosis: study protocol of a randomised phase IIa double-blinded placebo-controlled trial (SINAPPS2)', *Trials*, 20(1), 2019, pp.1–12

16 Sima, R., 'A catatonic woman awakened after 20 years. Her story may change psychiatry', *Washington Post*, 1 June 2023

17 Cullen, A. E., Holmes, S., Pollak, T. A. *et al.*, 'Associations between non-neurological autoimmune disorders and psychosis: a meta-analysis', *Biological Psychiatry*, 85(1), 2019, pp.35–48

18 Benros, M. E., Nielsen, P. R., Nordentoft, M. *et al.*, 'Autoimmune diseases and severe infections as risk factors for schizophrenia: a 30-year population-based register study', *American Journal of Psychiatry*, 168(12), 2011, pp.1303–10

19 Wang, L. Y., Chen, S. F., Chiang, J. H. *et al.*, 'Autoimmune diseases are associated with an increased risk of schizophrenia: a nationwide population-based cohort study', *Schizophrenia Research*, 202, 2018, pp.297–302

20 Benros, M. E., Pedersen, M. G., Rasmussen, H. *et al.*, 'A nationwide study on the risk of autoimmune diseases in individuals with a personal or a family history of schizophrenia and related psychosis', *American Journal of Psychiatry*, 171(2), 2014, pp.218–26

21 Davies, G., Welham, J., Chant, D. *et al.*, 'A systematic review and meta-analysis of Northern Hemisphere season of birth studies in schizophrenia', *Schizophrenia Bulletin*, 29(3), 2003, pp.587–93

22 Cheslack-Postava, K. and Brown, A. S., 'Prenatal infection and schizophrenia: a decade of further progress', *Schizophrenia Research*, 247, 2022, pp.7–15

23 Buckley, P. F., 'Neuroinflammation and schizophrenia', *Current Psychiatry Reports*, 21(8), 2019, pp.1–3

24 Miller, B. J., Buckley, P., Seabolt, W., 'Meta-analysis of cytokine alterations in schizophrenia: clinical status and antipsychotic effects', *Biological Psychiatry*, 70(7), 2011, pp.663–71

25 Galea, I., 'The blood–brain barrier in systemic infection and inflammation', *Cellular & Molecular Immunology*, 18(11), 2021, pp.2489–501

26 Armangue, T., Leypoldt, F., Málaga, I. *et al.*, 'Herpes simplex virus encephalitis is a trigger of brain autoimmunity, *Annals of Neurology*, 75(2), 2014, pp.317–23

27 Szeligowski, T., Yun, A. L., Lennox, B. R. *et al.*, 'The gut microbiome and schizophrenia: the current state of the field and clinical applications', *Frontiers in Psychiatry*, 11, 2020, p.156

28 International Schizophrenia Consortium, 'Common polygenic variation contributes to risk of schizophrenia and bipolar disorder', *Nature*, 460, 2009, pp.748–52

29 Avramopoulos, D., Pearce, B. D., McGrath, J. *et al.*, 'Infection and inflammation in schizophrenia and bipolar disorder: a genome wide study for interactions with genetic variation', *PLOS ONE*, 10(3), 2015, e0116696

30 Sekar, A., Bialas, A. R., De Rivera, H. *et al.*, 'Schizophrenia risk from complex variation of complement component 4', *Nature*, 530(7589), 2016, pp.177–83

31 Lataster, J., Myin-Germeys, I., Lieb, R. *et al.*, 'Adversity and psychosis: a 10-year prospective study investigating synergism between early and recent adversity in psychosis', *Acta Psychiatrica Scandinavica*, 125(5), 2012, pp.388–99

32 Song, H., Fang, F., Tomasson, G., *et al.*, 'Association of stress-related disorders with subsequent autoimmune disease', *JAMA*, 319(23), 2018, pp. 2388–400

7 The Inflamed Mind

1 Capuron, L. and Miller, A. H., 'Cytokines and psychopathology: lessons from interferon-α', *Biological Psychiatry*, 56(11), 2004, pp.819–24

2 National Institute for Health and Care Excellence (NICE), 'Depression: how common is it?', https://cks.nice.org.uk/topics/depression/background-information/prevalence/

3 Zimmerman, M., Ellison, W., Young, D. *et al.*, 'How many different ways do patients meet the diagnostic criteria for major depressive disorder?', *Comprehensive Psychiatry*, 56, 2015, pp.29–34

4 Warden, D., Rush, A. J., Trivedi, M. H. *et al.*, 'The STAR* D Project results: a comprehensive review of findings', *Current Psychiatry Reports*, 9(6), 2007, pp.449–59

5 Osimo, E. F., Baxter, L. J., Lewis, G. *et al.*, 'Prevalence of low-grade inflammation in depression: a systematic review and meta-analysis of CRP levels', *Psychological Medicine*, 49(12), 2019, pp.1958–70

6 Haapakoski, R., Mathieu, J., Ebmeier, K. P. *et al.*, 'Cumulative meta-analysis of interleukins 6 and 1β, tumour necrosis factor α and C-reactive protein in patients with major depressive disorder', *Brain, Behavior, and Immunity*, 49, 2015, pp.206–15

7 Foley, É. M., Parkinson, J. T., Mitchell, R. E. *et al.*, 'Peripheral blood cellular immunophenotype in depression: a systematic review and meta-analysis', *Molecular Psychiatry*, 2022, pp.1–16

8 Haroon, E., Daguanno, A. W., Woolwine, B. J. *et al.*, 'Antidepressant treatment resistance is associated with increased inflammatory markers in patients with major depressive disorder', *Psychoneuroendocrinology*, 95, 2018, pp.43–9; Arteaga-Henríquez, G., Simon, M. S., Burger, B. *et al.*, 'Low-grade inflammation as a predictor of antidepressant and anti-inflammatory therapy response in MDD patients: a systematic review of the literature in combination with an analysis of experimental data collected in the EU-MOODINFLAME consortium', *Frontiers in Psychiatry*, 10, 2019, p.458

9 *Spurious Correlations*, https://www.tylervigen.com/spurious-correlations

10 Wohleb, E. S., McKim, D. B., Sheridan, J. F. *et al.*, 'Monocyte trafficking to the brain with stress and inflammation: a novel axis of immune-to-brain communication that influences mood and behavior', *Frontiers in Neuroscience*, 8, 2015, p.447

11 Hodes, G. E., Pfau, M. L., Leboeuf, M. *et al.*, 'Individual differences in the peripheral immune system promote resilience versus susceptibility to social stress', *Proceedings of the National Academy of Sciences*, 111(45), 2014, pp.16136–41

12 Roberts, C., Sahakian, B. J. and Robbins, T. W., 'Psychological mechanisms and functions of 5-HT and SSRIs in potential therapeutic change: lessons from the serotonergic modulation of action selection, learning, affect, and social cognition', *Neuroscience & Biobehavioral Reviews*, 119, 2020, pp.138–67

13 Dantzer, R., 'Role of the kynurenine metabolism pathway in inflammation-induced depression: preclinical approaches', in Danzter, R. and Capuron, L. (eds), *Inflammation-Associated Depression: Evidence, Mechanisms and Implications*, Springer, 2016, pp.117–38

14 Hunt, C., Cordeiro, T. M. E., Suchting, R. *et al.*, 'Effect of immune activation on the kynurenine pathway and depression symptoms – a systematic review and meta-analysis', *Neuroscience & Biobehavioral Reviews*, 118, 2020, pp.514–23; Paul, E. R., Schwieler, L., Erhardt, S. *et al.*, 'Peripheral and central kynurenine pathway abnormalities in major depression', *Brain, Behavior, and Immunity*, 101, 2022, pp.136–45

15 Felger, J. C. and Treadway, M. T., 'Inflammation effects on motivation and motor activity: role of dopamine', *Neuropsychopharmacology*, 42(1), 2017, pp.216–41

16 Ekdahl, C. T., Kokaia, Z. and Lindvall, O., 'Brain inflammation and adult neurogenesis: the dual role of microglia', *Neuroscience*, 158(3), 2009, pp.1021–9

17 Vichaya, E. G., Malik, S., Sominsky, L. *et al.*, 'Microglia depletion fails to abrogate inflammation-induced sickness in mice and rats', *Journal of Neuroinflammation*, 17(1), 2020, pp.1–14

18 Harrison, N. A., Brydon, L., Walker, C. *et al.*, 'Neural origins of human sickness in interoceptive responses to inflammation', *Biological Psychiatry*, 66(5), 2009, pp.415–22

19 Brydon, L., Harrison, N. A., Walker, C. *et al.*, 'Peripheral inflammation is associated with altered substantia nigra activity and psychomotor slowing in humans', *Biological Psychiatry*, 63(11), pp.1022–9

20 Harrison, N. A., Voon, V., Cercignani, M. *et al.*, 'A neurocomputational account of how inflammation enhances sensitivity to punishments versus rewards', *Biological Psychiatry*, 80(1), 2016, pp.73–81

21 Davies, K. A., Cooper, E., Voon, V. *et al.*, 'Interferon and anti-TNF therapies differentially modulate amygdala reactivity which predicts associated bidirectional changes in depressive symptoms', *Molecular Psychiatry*, 26(9), 2021, pp.5150–60

22 Köhler-Forsberg, O., Lydholm, C. N., Hjorthøj, C. *et al.*, 'Efficacy of anti-inflammatory treatment on major depressive disorder or depressive symptoms: meta-analysis of clinical trials', *Acta Psychiatrica Scandinavica*, 139(5), 2019, pp.404–19; Kappelmann, N., Lewis, G., Dantzer, R. *et al.*, 'Antidepressant activity of anti-cytokine treatment: a systematic review and meta-analysis of clinical trials of chronic inflammatory conditions', *Molecular Psychiatry*, 23(2), 2018, pp.335–43

23 Khandaker, G. M., Pearson, R. M., Zammit, S., *et al.*, 'Association of serum interleukin 6 and C-reactive protein in childhood with depression and psychosis in young adult life: a population-based longitudinal study', *JAMA Psychiatry*, 71(10), 2014, pp.1121–8

24 Kappelmann, N., Arloth, J., Georgakis, M. K. *et al.*, 'Dissecting the association between inflammation, metabolic dysregulation, and specific depressive symptoms: a genetic correlation and 2-sample Mendelian randomization study', *JAMA Psychiatry*, 78(2), 2021, pp.161–70

25 Raison, C. L., Rutherford, R. E., Woolwine, B. J. *et al.*, 'A randomized controlled trial of the tumor necrosis factor antagonist infliximab for treatment-resistant depression: the role of baseline inflammatory biomarkers', *JAMA Psychiatry*, 70(1), 2013, pp.31–41

26 Salvadore, G., Nash, A., Bleys, C. *et al.*, 'A double-blind, placebo-controlled, multicenter study of Sirukumab as adjunctive treatment to a monoaminergic antidepressant in adults with major depressive disorder, in ACNP 57th annual meeting', *Neuropsychopharmacology*, 43(1), 2018, pp.228–382

27 O'Connor, J. C., Lawson, M. A., André, C. *et al.*, 'Lipopolysaccharide-induced depressive-like behavior is mediated by indoleamine 2, 3-dioxygenase activation in mice', *Molecular Psychiatry*, 14(5), 2009, pp.511–22

28 Dean, O. M., Kanchanatawan, B., Ashton, M. *et al.*, 'Adjunctive minocycline treatment for major depressive disorder: a proof of concept trial', *Australian & New Zealand Journal of Psychiatry*, 51(8), 2017, pp.829–40

29 Nettis, M. A., Lombardo, G., Hastings, C. *et al.*, 'Augmentation therapy with minocycline in treatment-resistant depression patients with low-grade

peripheral inflammation: results from a double-blind randomised clinical trial', *Neuropsychopharmacology*, 46(5), 2021, pp.939–48

30 Lynall, M. E., Turner, L., Bhatti, J. *et al.*, 'Peripheral blood cell–stratified subgroups of inflamed depression', *Biological Psychiatry*, 88(2), 2020, pp.185–96

31 Cattaneo, A., Ferrari, C., Turner, L. *et al.*, 'Whole-blood expression of inflammasome-and glucocorticoid-related mRNAs correctly separates treatment-resistant depressed patients from drug-free and responsive patients in the BIODEP study', *Translational Psychiatry*, 10(1), 2020, pp.1–14

32 van Eeden, W. A., van Hemert, A. M., Carlier, I. V. *et al.*, 'Basal and LPS-stimulated inflammatory markers and the course of individual symptoms of depression', *Translational Psychiatry*, 10(1), 2020, pp.1–12

33 Zwiep, J. C., Bet, P. M., Rhebergen, D. *et al.*, 'Efficacy of celecoxib add-on treatment for immuno-metabolic depression: protocol of the INFLAMED double-blind placebo-controlled randomized controlled trial', *Brain, Behavior, & Immunity-Health*, 2022, p.100585

34 Lamers, F., de Jonge, P., Nolen, W. A. *et al.*, 'Identifying depressive subtypes in a large cohort study: results from the Netherlands Study of Depression and Anxiety (NESDA)', *Journal of Clinical Psychiatry*, 71(12), 2010, p.8450

35 Milaneschi, Y., Lamers, F., Berk, M. *et al.*, 'Depression heterogeneity and its biological underpinnings: toward immunometabolic depression', *Biological Psychiatry*, 88(5), 2020, pp.369–80

36 Lamers, F., Milaneschi, Y., Smit, J. H. *et al.*, 'Longitudinal association between depression and inflammatory markers: results from the Netherlands study of depression and anxiety', *Biological Psychiatry*, 85(10), 2019, pp.829–37

37 Lynall, M. E., Soskic, B., Hayhurst, J. *et al.*, 'Genetic variants associated with psychiatric disorders are enriched at epigenetically active sites in lymphoid cells', *Nature Communications*, 13(1), 2022, pp.1–15

8 Inflammatory Thoughts

1 Koren, T., Amer, M., Krot, M. *et al.*, 'Insular cortex neurons encode and retrieve specific immune responses', *Cell*, 184(24), 2021, pp.5902–15

2 Koren, T. and Rolls, A., 'Immunoception: defining brain-regulated immunity', *Neuron*, 110(21), 2022, pp.3425–8

3 Mackenzie, J. N. 'The production of the so-called "rose cold" by means of an artificial rose, with remarks and historical notes', *American Journal of the Medical Sciences*, 91(181), 1886, p.45

4 Ader, R. and Cohen, N., 'Behaviorally conditioned immunosuppression', *Psychosomatic Medicine*, 37(4), 1975, pp.333–40

5 Hadamitzky, M., Lückemann, L., Pacheco-López, G. *et al.*, 'Pavlovian conditioning of immunological and neuroendocrine functions', *Physiological Reviews*, 100(1), pp.357–405

6 Lyman, M., *The Painful Truth: The New Science of Why We Hurt and How We Can Heal*, Bantam Press, 2021

7 Schaller, M., Miller, G. E., Gervais, W. M. *et al.*, 'Mere visual perception of other people's disease symptoms facilitates a more aggressive immune response', *Psychological Science*, 21(5), 2010, pp.649–52

8 Kayama, T., Ikegaya, Y. and Sasaki, T., 'Phasic firing of dopaminergic neurons in the ventral tegmental area triggers peripheral immune responses', *Scientific Reports*, 12(1), 2022, pp.1–9

9 Barlow, M. A., Wrosch, C., Gouin, J. P. *et al.*, 'Is anger, but not sadness, associated with chronic inflammation and illness in older adulthood?', *Psychology and Aging*, 34(3), 2019, p.330

10 Graham-Engeland, J. E., Sin, N. L., Smyth, J. M. *et al.*, 'Negative and positive affect as predictors of inflammation: timing matters', *Brain, Behavior, and Immunity*, 74, pp.222–30

11 Stellar, J. E., John-Henderson, N., Anderson, C. L. *et al.*, 'Positive affect and markers of inflammation: discrete positive emotions predict lower levels of inflammatory cytokines', *Emotion*, 15(2), 2015, p.129

12 Bierhaus, A., Wolf, J., Andrassy, M. *et al.*, 'A mechanism converting psychosocial stress into mononuclear cell activation', *Proceedings of the National Academy of Sciences*, 100(4), 2003, pp.1920–5

13 Dhabhar, F.S., Malarkey, W.B., Neri, E. and McEwen, B.S., 'Stress-induced redistribution of immune cells – from barracks to boulevards to battlefields: A tale of three hormones', Curt Richter Award Winner, *Psychoneuroendocrinology*, 37(9), 2012, pp.1345–68

14 Saint-Mezard, P., Chavagnac, C., Bosset, S. *et al.*, 'Psychological stress exerts an adjuvant effect on skin dendritic cell functions in vivo', *Journal of Immunology*, 171(8), 2003, pp.4073–80

15 Kataoka, N., Shima, Y., Nakajima, K. *et al.*, 'A central master driver of psychosocial stress responses in the rat', *Science*, 367(6482), 2020, pp.1105–12

16 Dhabhar, F. S., 'Effects of stress on immune function: the good, the bad, and the beautiful', *Immunologic Research*, 58(2), 2014, pp.193–210

17 Poller, W. C., Downey, J., Mooslechner, A. A. *et al.*, 'Brain motor and fear circuits regulate leukocytes during acute stress', *Nature*, 607(7919), 2022, pp.578–84

18 Haykin, H. and Rolls, A., 'The neuroimmune response during stress: a physiological perspective', *Immunity*, 54(9), 2021, pp.1933–47

19 Capellino, S., Claus, M. and Watzl, C., 'Regulation of natural killer cell activity by glucocorticoids, serotonin, dopamine, and epinephrine', *Cellular & Molecular Immunology*, 17(7), 2020, pp.705–11

20 Engler, H., Bailey, M. T., Engler, A. *et al.*, 'Effects of repeated social stress on leukocyte distribution in bone marrow, peripheral blood and spleen', *Journal of Neuroimmunology*, 148(1-2), 2004, pp.106–15

21 Hinwood, M., Morandini, J., Day, T. A. *et al.*, 'Evidence that microglia mediate the neurobiological effects of chronic psychological stress on the medial prefrontal cortex', *Cerebral Cortex*, 22(6), 2012, pp.1442–54

22 Lobo, B., Tramullas, M., Finger, B. C. *et al.*, 'The stressed gut: region-specific immune and neuroplasticity changes in response to chronic psychosocial stress', *Journal of Neurogastroenterology and Motility*, 29(1), 2023, pp.72–84

23 Patterson, A. M., Yildiz, V. O., Klatt, M. D. *et al.*, 'Perceived stress predicts allergy flares', *Annals of Allergy, Asthma & Immunology*, 112(4), 2014, pp.317–21

24 O'Donovan, A., Cohen, B. E., Seal, K. H. *et al.*, 'Elevated risk for autoimmune disorders in Iraq and Afghanistan veterans with posttraumatic stress disorder', *Biological Psychiatry*, 77(4), 2015, pp.365–74

25 Kuhlman, K. R., Cole, S. W., Irwin, M. R., *et al.*, 'The role of early life adversity and inflammation in stress-induced change in reward and risk processes among adolescents', *Brain, Behavior, and Immunity*, 109, 2023, pp.78–88

26 Nikkheslat, N., McLaughlin, A. P., Hastings, C. *et al.*, 'Childhood trauma, HPA axis activity and antidepressant response in patients with depression', *Brain, Behavior and Immunity*, 87, 2020, pp.229–37

27 Frank, M. G., Fonken, L. K., Watkins, L. R. *et al.*, 'Acute stress induces chronic neuroinflammatory, microglial and behavioral priming: a role for potentiated NLRP3 inflammasome activation', *Brain, Behavior, and Immunity*, 89, 2020, pp.32–42; Song, H., Fang, F., Tomasson, G. *et al.*, 'Association of stress-related disorders with subsequent autoimmune disease', *JAMA*, 319(23), 2018, pp.2388–400

28 Van Bogart, K., Engeland, C. G., Sliwinski, M. J. *et al.*, 'The association between loneliness and inflammation: findings from an older adult sample', *Frontiers in Behavioral Neuroscience*, 2022, p.360; Jaremka, L. M., Fagundes, C. P., Peng, J. *et al.*, 'Loneliness promotes inflammation during acute stress', *Psychological Science*, 24(7), 2013, pp.1089–97

29 Kuper, H. and Marmot, M., 'Job strain, job demands, decision latitude, and risk of coronary heart disease within the Whitehall II study', *Journal of Epidemiology & Community Health*, 57(2), 2003, pp.147–53; Väänänen, A., Koskinen, A., Joensuu, M. *et al.*, 'Lack of predictability at work and risk of acute myocardial infarction: an 18-year prospective study of industrial employees', *American Journal of Public Health*, 98(12), 2008, pp.2264–71

30 Biltz, R. G., Sawicki, C. M., Sheridan, J. F. *et al.*, 'The neuroimmunology of social-stress-induced sensitization', *Nature Immunology*, 2022, pp.1–9

9 Nobody's Land

1 Davis, H. E., Assaf, G. S., McCorkell, L. *et al.*, 'Characterizing long COVID in an international cohort: 7 months of symptoms and their impact', *EClinicalMedicine*, 38, 2021, 101019

2 Glaser, R., Padgett, D. A., Litsky, M. L. *et al.*, 'Stress-associated changes in the steady-state expression of latent Epstein–Barr virus: implications for chronic fatigue syndrome and cancer', *Brain, Behavior, and Immunity*, 19(2), 2005, pp.91–103

3 Meng, M., Zhang, S., Dong, X. *et al.*, 'COVID-19 associated EBV reactivation and effects of ganciclovir treatment', *Immunity, Inflammation and Disease*, 10(4), 2022, e597

4 Su, Y., Yuan, D., Chen, D. G., *et al.*, 'Multiple early factors anticipate post-acute COVID-19 sequelae', *Cell*, 185(5), 2022, pp.881–95

5 Klein, J., Wood, J., Jaycox, J. *et al.*, 'Distinguishing features of Long COVID identified through immune profiling', *medRxiv*, 2022, https://doi.org/10.1101/2022.08.09.22278592

6 Bjornevik, K., Cortese, M., Healy, B. C. *et al.*, 'Longitudinal analysis reveals high prevalence of Epstein-Barr virus associated with multiple sclerosis', *Science*, 375(6578), 2022, pp.296–301

7 'A study of an Epstein-Barr virus (EBV) candidate vaccine, mRNA-1189, in 18- to 30-year-old healthy adults', *Clinicaltrials.gov*, https://clinicaltrials.gov/ct2/show/NCT05164094

8 Cosentino, G., Todisco, M., Hota, N. *et al.*, 'Neuropathological findings from COVID-19 patients with neurological symptoms argue against a direct brain invasion of SARS-CoV-2: a critical systematic review', *European Journal of Neurology*, 28(11), 2021, pp.3856–65

9 Payus, A. O., Jeffree, M .S., Ohn, M. H. *et al.*, 'Immune-mediated neurological syndrome in SARS-CoV-2 infection: a review of literature on autoimmune encephalitis in COVID-19', *Neurological Sciences*, 43(3), 2022, pp.1533–47

10 Wang, E. Y., Mao, T., Klein, J. *et al.*, 'Diverse functional autoantibodies in patients with COVID-19', *Nature*, 595(7866), 2021, pp.283–8

11 Fernández-Castañeda, A., Lu, P., Geraghty, A. C. *et al.*, 'Mild respiratory COVID can cause multi-lineage neural cell and myelin dysregulation', *Cell*, 185(14), 2022, pp.2452–68

12 Yang, A. C., Kern, F., Losada, P. M. *et al.*, 'Dysregulation of brain and choroid plexus cell types in severe COVID-19', *Nature*, 595(7868), 2021, pp.565–71

13 Taquet, M., Sillett, R., Zhu, L. *et al.*, 'Neurological and psychiatric risk trajectories after SARS-CoV-2 infection: an analysis of 2-year retrospective cohort studies including 1 284 437 patients', *Lancet Psychiatry*, 9(10), 2022, pp.815–27

14 Ruffieux, H., Hanson, A. L., Lodge, S. *et al.*, 'A patient-centric modeling framework captures recovery from SARS-CoV-2 infection', *Nature Immunology*, 24, 2023, pp.349–58

15 Klein, J., Wood, J., Jaycox, J. *et al.*, 'Distinguishing features of Long COVID identified through immune profiling', *medRxiv*, 2022, https://doi.org/10.1101/2022.08.09.22278592

16 Su, Y., Yuan, D., Chen, D. G. *et al.*, 'Multiple early factors anticipate post-acute COVID-19 sequelae', *Cell*, 185(5), 2022, pp.881–95

17 Wang, S., Quan, L., Chavarro, J. E. *et al.*, 'Associations of depression, anxiety, worry, perceived stress, and loneliness prior to infection with risk of post–COVID-19 conditions', *JAMA Psychiatry*, 79(11), 2022, pp.1081–91

18 Liu, Q., Mak, J. W. Y., Su, Q. *et al.*, 'Gut microbiota dynamics in a prospective cohort of patients with post-acute COVID-19 syndrome', *Gut*, 71(3), 2022, pp.544–52

19 Thomas, R., Aldous, J. W. F., Forsyth, R. *et al.*, 'The influence of a blend of probiotic lactobacillus and prebiotic inulin on the duration and severity of symptoms among individuals with COVID-19', *Infectious Diseases Diagnosis and Treatment*, 5(1), 2021

20 Bramante, C. T., Buse, J. B., Liebovitz, D. *et al.*, 'Outpatient treatment of Covid-19 and incidence of post-Covid-19 condition over 10 months (COVID-OUT): a multicentre, randomised, quadruple-blind, parallel-group, phase 3 trial trial', *The Lancet Infectious Diseases*, 2023

21 Wu, H., Esteve, E., Tremaroli, V. *et al.*, 'Metformin alters the gut microbiome of individuals with treatment-naive type 2 diabetes, contributing to the therapeutic effects of the drug', *Nature Medicine*, 23(7), 2017, pp.850–8

22 Stephan, K. E., Manjaly, Z. M., Mathys, C. D. *et al.*, 'Allostatic self-efficacy: a metacognitive theory of dyshomeostasis-induced fatigue and depression', *Frontiers in Human Neuroscience*, 10, 2016, p.550

23 Peters, A., McEwen, B. S. and Friston, K., 'Uncertainty and stress: why it causes diseases and how it is mastered by the brain', *Progress in Neurobiology*, 156, 2017, pp.164–88

24 Kiverstein, J., Kirchhoff, M. D. and Thacker, M., 'An embodied predictive processing theory of pain experience', *Review of Philosophy and Psychology*, 13, 2022, pp.973–98

25 Ashar, Y. K., Gordon, A., Schubiner, H. *et al.*, 'Effect of pain reprocessing therapy vs placebo and usual care for patients with chronic back pain: a randomized clinical trial', *JAMA Psychiatry*, 79(1), 2022, pp.13–23

26 Kent, P., Haines, T., O'Sullivan, P., *et al.*, 'Cognitive functional therapy with or without movement sensor biofeedback versus usual care for chronic, disabling low back pain (RESTORE): a randomised, controlled, three-arm, parallel group, phase 3, clinical trial', *Lancet*, 401(10391), 2023, pp.1866–77

27 Vezzani, A. and Viviani, B., 'Neuromodulatory properties of inflammatory cytokines and their impact on neuronal excitability', *Neuropharmacology*, 96, 2015, pp.70–82

28 Sharp, H., Themelis, K., Amato, M. *et al.*, 'Fibromyalgia and myalgic encephalomyelitis/chronic fatigue syndrome (ME/CFS): an interoceptive predictive coding model of pain and fatigue expression, *Journal of Neurology, Neurosurgery & Psychiatry*, 92: A3–A4, 2021

10 The Cost of War

1 Tang, W., Kannaley, K., Friedman, D. B. *et al.*, 'Concern about developing Alzheimer's disease or dementia and intention to be screened: an analysis of national survey data', *Archives of Gerontology and Geriatrics*, 71, 2017, pp.43–9

2 Davis, D. H., Skelly, D. T., Murray, C. *et al.*, 'Worsening cognitive impairment and neurodegenerative pathology progressively increase risk for delirium', *American Journal of Geriatric Psychiatry*, 23(4), 2015, pp.403–15

3 Davis, D. H., Muniz-Terrera, G., Keage, H. A. *et al.*, 'Association of delirium with cognitive decline in late life: a neuropathologic study of 3 population-based cohort studies', *JAMA Psychiatry*, 74(3), 2017, pp.244–51

4 'Concerns grow over safety of Aduhelm after death of patient who got the drug', *New York Times*, 22 November 2021

5 Swardfager, W., Lanctôt, K., Rothenburg, L., *et al.*, 'A meta-analysis of cytokines in Alzheimer's disease', *Biological Psychiatry*, 68(10), 2010, pp. 930–41

6 King, E., O'Brien, J. T., Donaghy P. *et al.*, 'Peripheral inflammation in prodromal Alzheimer's and Lewy body dementias', *Journal of Neurology, Neurosurgery & Psychiatry*, 89(4), 2018, pp.339–45

7 Katan, M., Moon, Y. P., Paik, M. C. *et al.*, 'Infectious burden and cognitive function: the Northern Manhattan Study', *Neurology*, 80(13), pp.1209–15

8 Sipilä, P. N., Heikkilä, N., Lindbohm, J. V. *et al.*, 'Hospital-treated infectious diseases and the risk of dementia: a large, multicohort,

observational study with a replication cohort', *Lancet Infectious Diseases*, 21(11), 2021, pp.1557–67

9 Eyting, M., Xie, M., Heß, S. *et al.*, 'Causal evidence that herpes zoster vaccination prevents a proportion of dementia cases', *medRxiv*, 2023, https://doi.org/10.1101/2023.05.23.23290253

10 Lyman, M., Lloyd, D. G., Ji, X., *et al.*, 'Neuroinflammation: the role and consequences', *Neuroscience Research*, 79, 2014, pp.112; Yeung, C. H. C., Yeung, S. L. A. and Schooling, C. M., 'Association of autoimmune diseases with Alzheimer's disease: a mendelian randomization study', *Journal of Psychiatric Research*, 155, 2022, pp.550–8

11 Ide, M., Harris, M., Stevens, A. *et al.*, 'Periodontitis and cognitive decline in Alzheimer's disease', *PLOS ONE*, 11(3), 2016, e0151081

12 Kumar, D. K. V., Choi, S. H., Washicosky, K. J. *et al.*, 'Amyloid-β peptide protects against microbial infection in mouse and worm models of Alzheimer's disease', *Science Translational Medicine*, 8(340), 2016, 340ra72

13 Weaver, D. F., 'Alzheimer's disease as an innate autoimmune disease (AD²): a new molecular paradigm', *Alzheimer's & Dementia*, 19(3), 2022, pp.1086–98

14 Murray, C., Sanderson, D. J., Barkus, C. *et al.*, 'Systemic inflammation induces acute working memory deficits in the primed brain: relevance for delirium', *Neurobiology of Aging*, 33(3), 2012, pp.603–16

15 Wendeln, A. C., Degenhardt, K., Kaurani, L. *et al.*, 'Innate immune memory in the brain shapes neurological disease hallmarks', *Nature*, 556(7701), 2018, pp.332–8

16 Hong, S., Beja-Glasser, V. F., Nfonoyim, B. M. *et al.*, 'Complement and microglia mediate early synapse loss in Alzheimer mouse models', *Science*, 352(6286), 2016, pp.712–16

17 Shi, Q., Chowdhury, S., Ma, R. *et al.*, 'Complement C3 deficiency protects against neurodegeneration in aged plaque-rich APP/PS1 mice', *Science Translational Medicine*, 9(392), 2017, eaaf6295

18 Efthymiou, A. G. and Goate, A. M., 'Late onset Alzheimer's disease genetics implicates microglial pathways in disease risk', *Molecular Neurodegeneration*, 12(1), 2017, pp.1–12

19 Pascoal, T. A., Benedet, A. L., Ashton, N. J., Kang, M. S., Therriault, J., Chamoun, M., Savard, M., Lussier, F. Z., Tissot, C., Karikari, T. K. and Ottoy, J.; 'Microglial activation and tau propagate jointly across Braak stages', *Nature Medicine*, 27(9), 2021, pp.1592–99

20 Vogt, N. M., Kerby, R. L., Dill-McFarland, K. A., *et al.*, 'Gut microbiome alterations in Alzheimer's disease', *Scientific Reports*, 7(1), 2017, pp.1–11

21 Sochocka, M., Donskow-Łysoniewska, K., Diniz, B. S. *et al.*, 'The gut microbiome alterations and inflammation-driven pathogenesis of Alzheimer's disease – a critical review', *Molecular Neurobiology*, 56(3), 2019, pp.1841–51

22 Boehme, M., Guzzetta, K. E., Bastiaanssen, T. F. *et al.*, 'Microbiota from young mice counteracts selective age-associated behavioral deficits', *Nature Aging*, 1(8), 2021, pp.666–76

23 Bevan-Jones, W. R., Cope, T. E., Jones, P. S. *et al.*, 'Neuroinflammation and protein aggregation co-localize across the frontotemporal dementia spectrum'; *Brain*, 143(3), 2020, pp.1010–26

24 Soehnlein, O. and Libby, P., 'Targeting inflammation in atherosclerosis – from experimental insights to the clinic', *Nature Reviews Drug Discovery*, 20(8), 2021, pp.589–610

25 Roth, G. A., Mensah, G. A., Johnson, C. O. *et al.*, 'Global burden of cardiovascular diseases and risk factors, 1990–2019: update from the GBD 2019 study', *Journal of the American College of Cardiology*, 76(25), 2020, pp. 2982–3021

26 Furman, D., Campisi, J., Verdin, E. *et al.*, 'Chronic inflammation in the etiology of disease across the life span', *Nature Medicine*, 25(12), pp.1822–32

27 Franceschi, C., Garagnani, P., Parini, P. *et al.*, 'Inflammaging: a new immune–metabolic viewpoint for age-related diseases', *Nature Reviews Endocrinology*, 14(10), 2018, pp.576–90

28 Song, C., Shi, J., Zhang, P. *et al.*, 'Immunotherapy for Alzheimer's disease: targeting β-amyloid and beyond', *Translational Neurodegeneration*, 11(1), 2022, pp.1–17

29 Zhou, M., Xu, R., Kaelber, D. C. *et al.*, 'Tumor Necrosis Factor (TNF) blocking agents are associated with lower risk for Alzheimer's disease in

patients with rheumatoid arthritis and psoriasis', *PLOS ONE*, 15(3), 2020, e0229819

30 Decourt, B., Lahiri, D. K. and Sabbagh, M. N., 'Targeting tumor necrosis factor alpha for Alzheimer's disease', *Current Alzheimer Research*, 14(4), 2017, pp.412–25

Part Three: Resetting Your Defence System

11 The Anti-Inflammatory Life

1 Attia, P., *Outlive: The Science and Art of Longevity* (Vermillion, 2023)

12 Eat

1 Leeming, E. R., Johnson, A. J., Spector, T. D. *et al.*, 'Effect of diet on the gut microbiota: rethinking intervention duration', *Nutrients*, 11(12), 2019, p.2862; Singh, R. K., Chang, H. W., Yan, D. I. *et al.*, 'Influence of diet on the gut microbiome and implications for human health', *Journal of Translational Medicine*, 15(1), 2017, pp.1–17

2 Lozupone, C. A., Stombaugh, J. I., Gordon, J. I. *et al.*, 'Diversity, stability and resilience of the human gut microbiota', *Nature*, 489(7415), 2012, pp.220–30

3 Heiman, M. L. and Greenway, F. L., 'A healthy gastrointestinal microbiome is dependent on dietary diversity', *Molecular Metabolism*, 5(5), 2016, pp.317–20

4 Schnorr, S. L., Candela, M., Rampelli, S. *et al.*, 'Gut microbiome of the Hadza hunter-gatherers', *Nature Communications*, 5(1), 2014, p.3654

5 Deehan, E. C. and Walter, J., 'The fiber gap and the disappearing gut microbiome: implications for human nutrition', *Trends in Endocrinology & Metabolism*, 27(5), 2016, pp.239–42

6 Yatsunenko, T., Rey, F. E., Manary, M. J. *et al.*, 'Human gut microbiome viewed across age and geography', *Nature*, 486(7402), 2012, pp.222–7

7 Bolte, L. A., Vila, A. V., Imhann, F. *et al.*, 'Long-term dietary patterns are associated with pro-inflammatory and anti-inflammatory features of the gut microbiome', *Gut*, 70(7), 2021, pp.1287–8; Peirce, J. M. and Alviña, K., 'The role of inflammation and the gut microbiome in depression and anxiety', *Journal of Neuroscience Research*, 97(10), 2019, pp.1223–41

8 McDonald, D., Hyde, E., Debelius, J. W. *et al.*, 'American gut: an open platform for citizen science microbiome research', *Msystems*, 3(3), 2018, e00031-18

9 Wastyk, H. C., Fragiadakis, G. K., Perelman, D. *et al.*, 'Gut-microbiota-targeted diets modulate human immune status', *Cell*, 184(16), 2021, pp.4137-53

10 Menni, C., Louca, P., Berry, S. E. *et al.*, 'High intake of vegetables is linked to lower white blood cell profile and the effect is mediated by the gut microbiome', *BMC Medicine*, 19, 2021, pp.1–10

11 Wang, P., Song, M., Eliassen, A. H. *et al.*, 'Optimal dietary patterns for prevention of chronic disease', *Nature Medicine*, 2023, pp.1–10

12 Willett, W., Rockström, J., Loken, B. *et al.*, 'Food in the Anthropocene: the EAT–Lancet Commission on healthy diets from sustainable food systems', *Lancet*, 393(10170), 2019, pp.447–92

13 Metchnikoff, E., *The Prolongation of Life* (Putnam, 1907)

14 Berding, K., Bastiaanssen, T. F., Moloney, G. M. *et al.*, 'Feed your microbes to deal with stress: a psychobiotic diet impacts microbial stability and perceived stress in a healthy adult population', *Molecular Psychiatry*, 2022, pp.1–10

15 Pollan, M., *In Defence of Food* (Penguin, 2009)

13 Play

1 Dohnalová, L., Lundgren, P., Carty, J. R. *et al.*, 'A microbiome-dependent gut–brain pathway regulates motivation for exercise', *Nature*, 2022, pp.1–9

2 Wood, B. M., Harris, J. A., Raichlen, D .A. *et al.*, 'Gendered movement ecology and landscape use in Hadza hunter-gatherers', *Nature Human Behaviour*, 5(4), 2021, pp.436–46

3 Arundell, L., Fletcher, E., Salmon, J. *et al.*, 'A systematic review of the prevalence of sedentary behavior during the after-school period among children aged 5-18 years', *International Journal of Behavioral Nutrition and Physical Activity*, 13(1), 2016, pp.1–9

4 Warburton, D. E. and Bredin, S. S., 'Health benefits of physical activity: a systematic review of current systematic reviews', *Current Opinion in Cardiology*, 32(5), 2017, pp.541–56

5 Takács, E., Barkaszi, I., Czigler, I. *et al.*, 'Persistent deterioration of visuospatial performance in spaceflight', *Scientific Reports*, 11(1), 2021, pp.1–11; Haapala, E. A., Väistö, J., Lintu, N., Westgate, K., Ekelund, U., Poikkeus, A. M., Brage, S. and Lakka, T. A., 'Physical activity and sedentary time in relation to academic achievement in children', *Journal of Science and Medicine in Sport*, 20(6), 2017, pp.583–9

6 Raichlen, D. A. and Alexander, G. E., 'Adaptive capacity: an evolutionary neuroscience model linking exercise, cognition, and brain health', *Trends in Neurosciences*, 40(7), 2017, pp.408–21

7 Miles, L. K., Karpinska, K., Lumsden, J. *et al.*, 'The meandering mind: vection and mental time travel', *PLOS ONE*, 5(5), 2010, e10825

8 Noetel, M., Sanders, T., Gallardo Gómez, D. *et al.*, 'Running from depression: a systematic review and network meta-analysis of exercise dose and modality in the treatment for depression', *SSRN*, http://dx.doi.org/10.2139/ssrn.4388153

9 Zaccardi, F., Davies, M. J., Khunti, K. *et al.*, 'Comparative relevance of physical fitness and adiposity on life expectancy: a UK Biobank observational study', *Mayo Clinic Proceedings*, 94(6), 2019, pp.985–94; Gaesser, G. A. and Angadi, S. S., 'Obesity treatment: Weight loss versus increasing fitness and physical activity for reducing health risks', *Iscience*, 24(10), 2021, p.102995

10 Falconer, C. L., Cooper, A. R., Walhin, J. P. *et al.*, 'Sedentary time and markers of inflammation in people with newly diagnosed type 2 diabetes', *Nutrition, Metabolism and Cardiovascular Diseases*, 24(9), 2014, pp.956–62

11 Phillips, C. M., Dillon, C. B. and Perry, I .J., 'Does replacing sedentary behaviour with light or moderate to vigorous physical activity modulate inflammatory status in adults?', *International Journal of Behavioral Nutrition and Physical Activity*, 14(1), 2017, pp.1–12

12 Nieman, D. C., Henson, D. A., Austin, M. D. *et al.*, 'Immune response to a 30-minute walk', *Medicine & Science in Sports & Exercise*, 37(1), 2005, pp.57–62

13 Shen, B., Tasdogan, A., Ubellacker, J. M., *et al.*, 'A mechanosensitive peri-arteriolar niche for osteogenesis and lymphopoiesis', *Nature*, 591(7850), 2021, pp.438–44

14 Dimitrov, S., Hulteng, E. and Hong, S., 'Inflammation and exercise: inhibition of monocytic intracellular TNF production by acute exercise via β2-adrenergic activation', *Brain, Behavior, and Immunity*, 61, 2017, pp.60–68

15 Duggal, N. A., Pollock, R. D., Lazarus, N. R. *et al.*, 'Major features of immunesenescence, including reduced thymic output, are ameliorated by high levels of physical activity in adulthood', *Aging Cell*, 17(2), 2018, e12750

16 Nelke, C., Dziewas, R., Minnerup, J. *et al.*, 'Skeletal muscle as potential central link between sarcopenia and immune senescence', *EBioMedicine*, 49, 2019, pp.381–8

17 Reilly, S. M. and Saltiel, A. R. 'Adapting to obesity with adipose tissue inflammation', *Nature Reviews Endocrinology*, 13(11), 2017, pp.633–43

18 Mailing, L. J., Allen, J. M., Buford, T.W. *et al.*, 'Exercise and the gut microbiome: a review of the evidence, potential mechanisms, and implications for human health', *Exercise and Sport Sciences Reviews*, 47(2), 2019, pp.75–85

19 White, M. P., Alcock, I., Grellier, J. *et al.*, 'Spending at least 120 minutes a week in nature is associated with good health and wellbeing', *Scientific Reports*, 9(1), 2019, pp.1–11

20 Thompson, C. W., Roe, J., Aspinall, P. *et al.*, 'More green space is linked to less stress in deprived communities: evidence from salivary cortisol patterns', *Landscape and Urban Planning*, 105(3), 2012, pp.221–9

21 Beyer, K. M., Kaltenbach, A., Szabo, A., 'Exposure to neighborhood green space and mental health: evidence from the survey of the health of Wisconsin', *International Journal of Environmental Research and Public Health*, 11(3), 2014, pp.3453–72

22 Jimenez, M. P., Elliott, E. G., DeVille, N.V. *et al.*, 'Residential green space and cognitive function in a large cohort of middle-aged women', *JAMA Network Open*, 5(4), 2022, e229306

23 Kuo, M., 'How might contact with nature promote human health? Promising mechanisms and a possible central pathway', *Frontiers in Psychology*, 6, 2015, p.1093

24 Selway, C. A., Mills, J. G., Weinstein, P. *et al.*, 'Transfer of environmental microbes to the skin and respiratory tract of humans after urban green space exposure', *Environment International*, 145, 2020, p.106084

25 Walker, M., *Why We Sleep: Unlocking the Power of Sleep and Dreams* (Simon & Schuster, 2017)

26 Prather, A. A., Janicki-Deverts, D., Hall, M. H. *et al.*, 'Behaviorally assessed sleep and susceptibility to the common cold', *Sleep*, 38(9), 2015, pp.1353–9

27 Spiegel, K., Sheridan, J. F. and Van Cauter, E., 'Effect of sleep deprivation on response to immunizaton', *JAMA*, 288(12), 2002, pp.1471–2

28 Suzuki, H., Savitz, J., Teague, T. K. *et al.*, 'Altered populations of natural killer cells, cytotoxic T lymphocytes, and regulatory T cells in major depressive disorder: association with sleep disturbance', *Brain, Behavior and Immunity*, 66, 2017, pp.193–200

29 Irwin, M., Mascovich, A., Gillin, J. C., 'Partial sleep deprivation reduces natural killer cell activity in humans', *Psychosomatic Medicine*, 56(6), 1994, pp.493–8

30 Irwin, M. R., Wang, M., Ribeiro, D. *et al.*, 'Sleep loss activates cellular inflammatory signaling', *Biological Psychiatry*, 64(6), 2008, pp.538–40; Irwin, M. R., 'Sleep and inflammation: partners in sickness and in health', *Nature Reviews Immunology*, 19(11), 2019, pp.702–15

31 Goldstein, A. N. and Walker, M. P., 'The role of sleep in emotional brain function', *Annual Review of Clinical Psychology*, 10, 2014, pp.679–708; Klinzing, J. G., Niethard, N. and Born, J., 'Mechanisms of systems memory consolidation during sleep', *Nature Neuroscience*, 22(10), 2019, pp.1598–1610

32 Smith, R. P., Easson, C., Lyle, S. M. *et al.*, 'Gut microbiome diversity is associated with sleep physiology in humans', *PLOS ONE*, 14(10), 2019, e0222394

33 Leproult, R., Colecchia, E. F., L'Hermite-Balériaux, M. *et al.*, 'Transition from dim to bright light in the morning induces an immediate elevation of cortisol levels', *Journal of Clinical Endocrinology & Metabolism*, 86(1), 2001, pp.151–7

34 Angarita, G. A., Emadi, N., Hodges, S. *et al.*, 'Sleep abnormalities associated with alcohol, cannabis, cocaine, and opiate use: a comprehensive review', *Addiction Science & Clinical Practice*, 11(1), 2016, pp.1–17

14 Love

1 Dantzer, R., Cohen, S., Russo, S. J. *et al.*, 'Resilience and immunity', *Brain, Behavior, and Immunity*, 74, 2018, pp.28–42

2 https://thinkcbt.com/think-cbt-worksheets

3 Starr, L. R., Hershenberg, R., Shaw, Z. A. *et al.*, 'The perils of murky emotions: emotion differentiation moderates the prospective relationship between naturalistic stress exposure and adolescent depression', *Emotion*, 20(6), 2020, p.927

4 Hoemann, K., Xu, F. and Barrett, L. F., 'Emotion words, emotion concepts, and emotional development in children: A constructionist hypothesis', *Developmental Psychology*, 55(9), 2019, p.1830

5 Willcox, G., 'The feeling wheel: a tool for expanding awareness of emotions and increasing spontaneity and intimacy', *Transactional Analysis Journal*, 12(4), 1982, pp.274–6

6 Hazlett, L. I., Moieni, M., Irwin, M. R. *et al.*, 'Exploring neural mechanisms of the health benefits of gratitude in women: a randomized controlled trial', *Brain, Behavior, and Immunity*, 95, 2021, pp.444–53

7 Hölzel, B. K., Carmody, J., Vangel, M. *et al.*, 'Mindfulness practice leads to increases in regional brain gray matter density', *Psychiatry Research: Neuroimaging*, 191(1), 2011, pp.36–43

8 Stellar, J. E., John-Henderson, N., Anderson, C. L. *et al.*, 'Positive affect and markers of inflammation: discrete positive emotions predict lower levels of inflammatory cytokines', *Emotion*, 15(2), 2015, p.129

9 'Tai chi: what you need to know', *National Centre for Complementary and Integrative Health*, https://www.nccih.nih.gov/health/tai-chi-what-you-need -to-know#:~:text=Tai%20chi%20is%20a%20practice,on%20health%20 promotion%20and%20rehabilitation

10 Sapolsky, R .M., *Why Zebras Don't Get Ulcers: The Acclaimed Guide to Stress, Stress-Related Diseases, and Coping* (Holt Paperbacks, 2004)

11 Sapolsky, R. M., 'Why Zebras Don't Get Ulcers: Stress and Health', lecture given to the Beckman Institute, University of Illinois Urbana-Champaign, 20 June 2017

12 Dweck, C. S. and Yeager, D. S., 'Mindsets: a view from two eras', *Perspectives on Psychological Science*, 14(3), 2019, pp.481–96

13 Durant, W., *Story of Philosophy* (Simon & Schuster, 1961)

14 Clear, J., *Atomic Habits: An Easy & Proven Way to Build Good Habits & Break Bad Ones* (Penguin, 2018)

15 Holt-Lunstad, J., 'Social connection as a public health issue: the evidence and a systemic framework for prioritizing the "social" in social determinants of health', *Annual Review of Public Health*, 43, 2022, pp.193–213

16 Eisenberger, N. I., Lieberman, M. D. and Williams, K. D., 'Does rejection hurt? An fMRI study of social exclusion', *Science*, 302(5643), 2003, pp.290–2

17 Sapolsky, R. M., 'Social status and health in humans and other animals', *Annual Review of Anthropology*, 33, 2004, pp.393–418

Acknowledgments

There are many people without whom this book would never have made it into your hands. I would like to place two particular groups of brave people in the vanguard, and after that there is no particular order of precedence.

The first is the patients. Individuals living with mental health conditions are among the most misunderstood and stigmatized in society. This book is filled with excitement about new discoveries of immune-brain pathways, newly discovered conditions and novel therapies. This means, however, that until recently, most of those living with mental symptoms caused or worsened by the immune system have been misdiagnosed and mistreated. I would like to thank my patients, who have taught me so much about the human condition, and have inspired a number of cases in this book. I'd also love to thank those who are not my patients but were willing to share their stories of illness, Samantha Raggio being the only one I can name.

The second group is the pioneers. These are the scientists and doctors who have fought against convention, professional ridicule and even persecution to reveal the links between the mind, the immune system and the microbiome. This book is buttressed by around 350 references, each representing a group of pioneering researchers. I would like to thank the experts who kindly let me interview them or reviewed the book: Belinda Lennox, Golam Khandaker, John Cryan, Brenda Penninx, Jonathan Kipnis, David Zilber, Iain McInnes, Ed Bullmore and Awais Aftab.

To my editor: the skilful and wise Alex Christofi. And to my wonderful team at Transworld: Tom Hill, Alex Newby, Kate Samano,

Phil Lord, Rich Shailer, Eloise Austin and Phil Evans. You have all helped make an accessible and enjoyable book out of what must be one of the most complicated corners of science.

To my magnificent agent, Charlie Viney. I cannot believe that this is my third book.

To colleagues and advisors who have advised and inspired me in the realms of immunology and psychiatry: Amedeo Minichino, Olga Tsatalou, Jakov Zlodre, Lilian Hickey, Daniel Maughan, Jonathan Rogers, Margaret Glogowska, Graham Ogg, Rebecca McKnight, Kate Thomas, Dafydd Lloyd and John Beale.

To those who have supported my literary career: Colin Thubron, Margreta de Grazia, Andrea Henry and Doug Young.

To Hannah, my wife and 'other editor'. Your mix of encouragement and honesty are perfect for both roles.

To my parents, Rob and Hannah, and my brother, Phin. Your patience in waiting three books for a dedication reflects your unceasing patience with having to put up with me for a son/sibling . . .

And – of course – to you, the reader. I hope that you've enjoyed *The Immune Mind*, and I hope it gives you a different, deeper perspective on what it means to be human.

Index

5-aminovaleric acid (5AV) 86

Adelson's checker-shadow illusion 51, 52
adenosine 215–16
Ader, Robert 61–2
adjuvants 149
adrenaline 140, 148
aducanumab 173
adverse childhood experiences (ACEs) 153
alcohol 19, 216–17, 219
allergies 66, 82–3, 143, 152–3
Alzheimer, Alois 172
Alzheimer's disease
  amyloid plaques 172–3, 176
  as autoimmune disease 175
  hippocampus, impact on 121–2
  medication research 173, 179–80
  microglia links 175–6
  preventing 176, 177
  risk factors
    chronic inflammation 174–6
    dysbiosis 177
    genetic 173, 176
    infections 174–5
  symptoms 172
  tau tangles 172–3
American Gut Project 192
amygdala 75, 127, 147–8, 224
amyloid beta protein 172, 173, 175, 176
amyloid plaques 172–3, 175, 176
analogies see metaphors and analogies
anger 144–5
anhedonia 41, 117, 118, 132, 134
animal welfare 33n
anthrax 12
anti-inflammatories
  biologics 114–15, 128, 132–3
  INFLAMED trial 134–5
  psychological impacts 115, 127, 128–9,
    131–3
  steroids 99, 101

anti-NMDA receptor encephalitis
  cases
    Cahalan, Susannah 100–1
    Dr Dalmau's patients 98–9
    Raggio, Samantha 93–7
  diagnosing difficulties 103
  discovery of 98–100
  infection link 109, 167
  Lennox's studies 102, 103–4
  misdiagnoses 94, 96, 101, 103, 104
  mode of action 99–100
  symptoms 98, 102–3
  treatments 99, 101, 103–4, 106
  undiagnosed cases 103
antibiotics, pharmaceutical 72, 76, 78, 132–3, 191
antibodies
  B-cell production of 17
  brain-attacking 59, 99, 101, 102, 103–4, 106–7,
    108
  discovery of 12
  lab-produced 114–15, 128–9, 132–3, 173
  mode of action 17
  virus impacts 161, 162
antidepressants 114–15, 117, 118, 128, 132–3
antigens 16, 27, 30, 62
anxiety 38, 65, 75, 85, 212
APOE4 gene 173
arachnoid mater 23
Aristotle 5, 229
arthritis 65, 114–15, 127, 128, 179
asthma 128
astrocytes 26, 29
atherosclerosis 178
atherosclerotic diseases 178, 186
Atomic Habits (Clear) 229–30
autism spectrum disorder (ASD) 85–6
autoimmune diseases 65, 66, 107–9, 110, 111, 114,
    132, 136, 152–3, 160, 162, 179; see also
    specific disease
autoimmune psychosis 103–9, 112; see also
    anti-NMDA receptor encephalitis

autoimmunity 153, 162
autonomic nervous system 7
avoidance, disease/infection 41–7, 62; *see also* sickness behaviour
Avon Longitudinal Study of Parents and Children (ALSPAC) study 129–30, 131
awe 89–90, 225

B-cells 17, 160
Bacillus anthracis 12
bacteria 9, 71–2, 73; *see also* gut microbiome; microbiome; pathogens; probiotics
Barrett, Lisa Feldman 55–6
behaviour; *see also* mental health/illness
    conditioning 61–2, 75, 143–4; *see also* predictive processing (coding)
    dementia impacts 177–8
    disease/infection avoidance 41–7, 62
    immune system influencing 36–7; *see also* sickness behaviour
    microbial manipulation of
        by gut microbes 74–7, 80–1, 82, 85–6, 205–6
        by pathogens 68–9, 79–80
behavioural activation 201
Bell, Charles 6
Bifidobacteria infantis 73
biologics 114–15, 128, 132–3
biomarkers 133–4
Bir Tawil 156
Black Death 65
blame culture 228
blood-brain barrier 19, 29, 30, 32, 33, 109, 119, 132–3
bone marrow 28, 113, 140, 148, 149, 152, 209
Boston Children's Hospital, MA, USA 31
Bouchard, Charles 71
Boyle, Robert 4, 9
brain; *see also* defence supersystem; nervous system; *specific cells*; *specific structures*
    abnormalities, structural 75–6
    age-related changes 30–1
    antibody attacks 59, 99, 101, 102, 103–4, 106–7, 108
    blood-brain barrier 19, 29, 30, 32, 33, 109, 119, 132–3
    brain lactate 84
    cerebral energy crisis 166

chronic inflammation impacts 120, 152, 153–4, 161, 165–6; *see also* Alzheimer's disease; dementia; neuroinflammation; psychosis
chronic pain 166–7
cleaning system 26–7
cognitive behavioural therapy (CBT) 201
cognitive dysfunction 162–3; *see also specific disorder*
cognitive functional therapy (CFT) 166–7
as command centre 8
drainage of 23, 25–7
emotions *see* emotions
growth 84
gut-brain communication 81–2
immune-brain loop 32–4
immune cells analysing waste 27
immune privilege 19, 22
inflammation of *see* encephalitis, autoimmune; neuroinflammation
location of the mind 5, 6, 18
memory of illness/inflammation 142
meningeal immune system 23–8
mental health *see* mental health/illness
neurodegeneration *see* neurodegeneration
neurodevelopmental conditions 31–2
neurogenesis 121–2
neuroplasticity 30–1
pain preprocessing therapy 166
perception 50–60, 66–7, 100, 165; *see also* delusions; hallucinations
post-viral illnesses impacting 162–3
predictive processing 50–7, 61–3, 66–7, 165–7, 207
reward processing 126, 134, 153
scientist metaphor 53, 55
stress response *see* stress response
threat detection 7–8, 17, 20, 28, 60–1, 127, 143, 144
brain-derived neurotrophic factor (BDNF), 73
brain lactate 84
Brain on Fire (Cahalan) 101, 103
breast milk, human 83–4, 85, 189
Bristol Medical School, England, UK 129
Bullmore, Ed 118
Burrell, April 107–8

C-reactive protein (CRP) 118, 129–30, 131, 132, 133
Caesarean section (C-section) births 82–3

caffeine 19, 216, 219
Cahalan, Susannah 100–1, 103
California Institute of Technology, Pasadena, CA, USA 86
cancer 19, 116, 151, 160, 187, 195, 209, 214–15
cannabinoid receptor 1 (CB1) 205
cannabis 72
Cardiff University, Wales, UK 122–7
Carroll, Lewis 204
Cartesian dualism 18
causation versus correlation 118–19
celecoxib 134–5
'cell,' first use of term 9
central nervous system (CNS) 7; *see also specific part*
cerebral cortex 5
cerebral energy crisis 166
Cerebri Anatome (Willis) 5–6, 17
cerebrospinal fluid (CSF) 23, 25–7, 28, 29, 32, 98
chemogenetics 34, 142
chickenpox 151–2
cholera 12, 71
Christian Hansen company 197–8
chromosome 21 173
chronic inflammation; *see also* inflammation/inflammatory response
  anger influencing 144–5
  atherosclerosis, causing 178
  brain, impact on 120, 152, 153–4, 161, 165–6; *see also* Alzheimer's disease; dementia; neuroinflammation; psychosis
  drug impacts *see* anti-inflammatories
  gut microbiome links 152, 177
  long-term sickness behaviour, causing 120, 161
  metabolic syndrome 135–6
  movement, reducing through 208–10
  root of mental and physical illnesses 178–9, 186–7
  sleep impacts 215
  socioeconomic links 231–2
  stress links 66, 152, 153–4, 226
  whole-body impacts 111
chronic stress 150–5, 226
circadian rhythm 216
circumventricular organs 32
cirrhosis of the liver 116
Clear, James 229–30
clinical depression *see* depression, clinical
Clostridioides difficile 43

Clostridium difficile 78
coeliac disease 108
cognitive behavioural therapy (CBT) 201
cognitive dysfunction 162–3; *see also specific disorder*
cognitive function, lifestyle factors impacting 207, 215
cognitive functional therapy (CFT) 166–7
complement proteins 31–2, 176
conditioning, behavioural 61–2, 75, 143–4
confirmation bias 56
control, sense of 154, 165–6, 213, 222, 226–7
correlation versus causation 118–19
corticosterone 73
cortisol 148, 150, 153, 163, 216
Covid-19 42, 44, 47, 58, 157, 158–61, 162–4
Cowper, William 188
cowpox 11
cows 189
crenezumab 173
Crohn's disease 65
Cryan, John 73–5, 77, 85, 86, 87–8, 89, 177, 200
cytokines; *see also specific cytokine*
  Alzheimer's link 174
  depression link 120–1, 127, 133
  drugs targeting *see* anti-inflammatories
  neuro-immune communication role 22, 32, 33, 40, 82, 119
  schizophrenia link 109
  secretion by macrophages 15
  secretion by microglia 30, 120
  stress response 144, 148–9
  triggering inflammatory response 15, 30

Dalmau, Josep 98–100, 101
dance 208
defence supersystem; *see also* brain; gut microbiome; immune system; nervous system; sickness behaviour; stress response
  balance/homeostasis 33, 54, 56, 81, 145, 186, 187
  bowel inflammation studies 33–4
  car analogy 180
  communication 22, 27, 28, 30, 32–4, 63–4, 74–5, 81–2, 119–20, 125–6, 142
  disease/infection avoidance 41–8, 62
  hypersensitivity 57, 64, 66, 109, 143, 147–8, 154
  immunoception 57–60
  inflammatory reflex 33

defence supersystem – *cont.*
   interconnectivity 22, 33–4, 64, 81–2, 119,
       125–6, 138, 144, 154–5, 165, 168–9, 183–4,
       185–7, 212, 220–1
   memory of illness/inflammation 13, 14, 17,
       34, 39, 62, 142, 221
   Ministry of Defence metaphor, The 63–4,
       185–6
   modern imbalance of 64–5, 66
   predictive processing 61–3, 165–7
   redundancy 125
   resetting *see* diet for defence supersystem;
       loving for defence supersystem;
       movement for defence supersystem
   Schwartz's discoveries 21–2
   separateness belief 17, 18, 19–20, 22, 24, 34,
       49–50, 143, 221
   sickness behaviour *see* sickness behaviour
   threat detection 60–1, 63–4, 144, 147, 185–6
   triangle metaphor 168
   unified defence system 60–4
   wheel analogy 138, 186–7, 210
   whole-body involvement 60–4, 140–1,
       168–9
Del Río Hortega, Pío 29
delirium 171–2
delusions 56, 96; *see also* anti-NMDA receptor
   encephalitis
dementia; *see also* Alzheimer's disease
   delirium links 171–2
   dysbiosis link 177
   frontotemporal 177–8
   infections increasing risk of 174–5
   inflammation links 171–2, 174–5, 178, 180,
       187
   preventing 176
   treatments, hopes for 179–80
   vascular 178
depression, clinical
   defining 116–17
   diverse causes, mechanisms and
       presentations 117–18
   drug impacts on
       anti-inflammatories 115, 127, 128–9, 131–3,
           134–5
       biologics 115, 128, 132–3
       IFN-α 116
       pro-inflammatories 116, 126–7
   genetic links 131
   immunometabolic depression 135–6

inflammation links
   ALSPAC study 129–30, 131
   anti-inflammatories relieving 115, 127, 128,
       131–3
   correlation versus causation 118–19
   cytokine impacts 120–1, 127
   diagnosing 133–4, 134–5
   endotoxin studies 38–9
   inflamed depression 118
   INFLAMED trial 134–5
   long Covid 163
   metabolism disturbances 135–6
   NESDA study 135–6
   neurogenesis, impact on 122
   pro-inflammatories increasing 116,
       126–7
   raised inflammatory marker levels 118
   reward and punishment sensitivity, impact
       on 126, 134
   sickness behaviour 119–21, 122–3, 123–5,
       126–7
   symptoms 133–4, 135
   trauma impacts 153
   two-way directionality 136, 137
lifestyle links 136, 137
metabolism links 135–6
microbiome links 77, 84
neurogenesis links 121–2
prevalence 117
stratification 118
symptoms 116–17, 118, 135
treatments
   anti-inflammatories 115, 127, 128–9, 131–3,
       134–5
   antidepressants 117
   behavioural activation 201, 207–8
   biologics 115, 128, 132–3
   lifestyle changes 136
   psychological interventions 117
whole-body disorder 137
Descartes, René 18, 20, 49
Deter, Auguste 172
diabetes mellitus 43, 136, 179, 187, 195
diet for defence supersystem
   80:20 rule 202
   behaviour, impacts on 205–6
   diversity 190, 191, 192, 193–4, 201, 202
   'eat the rainbow' 195, 203
   fermented foods 196–200, 203
   gut gardening 200–3, 233

gut microbe diversity, improving 191, 192, 194–5, 196
gut rehabilitation 193–5
key takeaways 202–3
making changes slowly 193–5, 203
mindfulness 233
planetary benefits 195–6
plant fibre 190–6, 202
psychobiotic diet 200
and sleep 216–17
ultra-processed foods, limiting 203
vitamins and phytochemicals 195
diet, planetary health 195–6
diffusion-weighted imaging MRI scan 125
diphtheria 12
disgust 44, 46–7
dissociation 56
dopamine 82, 121, 126, 144, 205
dura mater 23, 24–5, 26, 27, 28
dural sinuses 23, 24–5, 26, 27
Durant, Will 229
dysbiosis 83, 86, 89, 109–10, 164, 177

eczema 153, 194, 215
Ehrlich, Paul 12
Eliot, T. S. 170
Emory University, Atlanta, GA, USA 132
emotional regulation 145
emotions 55–6, 75, 122, 144–6, 221, 223–4; see also disgust; mood
encephalitis, autoimmune 102–3, 162; see also anti-NMDA receptor encephalitis
endotoxin (lipopolysaccharide (LPS)) 38–9, 43, 122–3, 123–5
enteric nervous system 81
enteroendocrine cells 82
ependymal cells 29
epidemiology 129
Epstein-Barr virus (EBV) 160–2
Escheria coli 73
eustress 227
exercise *see* movement for defence supersystem
exteroception 55

faecal transplants 76–7, 87–8, 177, 180
fatigue, persistent 157–8, 166, 167
fatty acid amides (FAAs) 205
fear of contagion 42, 44–8
fermented foods 70, 71, 90, 196–200, 201, 203
fibre, plant 189, 190–6, 200, 201, 202

fight-or-flight-or-freeze response 75, 140, 147–8
Fleming, Alexander 72
flexibility 184, 202, 227–8
food preferences 80, 81, 82
four horsemen of chronic disease 186–7
Frankl, Viktor 145–6, 229
Friston, Karl 54, 62, 63, 169
frontotemporal dementia 177–8

Galen 5
gamma-aminobutyric acid (GABA) 82, 86
gantenerumab 173
gastrointestinal tract 78; see also gut microbiome
genetics 65, 77, 110, 111, 130–1, 173, 176
germ-free mice 72–3, 74–6, 76–7, 80, 85
germ theory 12, 70–1, 71–2
Glasgow Royal Infirmary, Scotland, UK 114
glial cells 26, 29, 32; see also microglia
glucocorticoid resistance 153
glutamate 99
glymphatic system 26–7, 32
Good, Sarah 105
gratitude practice 224
Great Plague of London 10
gut microbiome; see also defence supersystem
ageing, impact on 87–8
antibiotic impacts on 78
behaviour, impact on 73, 74–7, 80–1, 82, 85–6
breast-feeding impacts 83–4, 189
chronic inflammation impacts 152, 177
defence against pathogens 78–9
and dementia 177
development 84–5
digestion, help with 188–9
diversity 190, 191
dysbiosis 83, 86, 89, 109–10, 177
fibre for, importance of 189
food preferences, impact on 80, 82
gut-brain communication 81–2
gut gardening 200–3, 233
gut rehabilitation 193–5
as immune organ 78–9, 88
immune system links 82
improving 71; see also diet for defence supersystem
inflammation links 78–9, 86, 87, 89, 195, 196
mental health links 74–7, 84–6, 88–9
Metchnikoff's research 70–1

gut microbiome – *cont.*
  movement links 204–6, 210
  neurodegenerative diseases, impacts on
      87–8
  number of microbes 188
  pet-ownership impacts 212
  present at birth 82–3
  psychosis links 109–10
  role 188–9
  SCFA production 78, 82
  sleep impacts 215
  stability 78
  stress links 74–5, 111, 152
  threat detection 75
  time in nature, impacts of 212
  treatments for post-viral illnesses 164

H1N1 influenza (Spanish flu) 72
habits 211, 229–30
Hadza people, Tanzania 191, 205
hallucinations 56, 96, 107–8; *see also* anti-
      NMDA receptor encephalitis
hallucinogenic drugs 100
Harrison, Neil 122–3, 125–7, 134, 142
Harvard University, Cambridge, MA, USA 28,
      149, 161–2, 176, 195
heart disease 136, 186, 195
help, asking for 221–2
helplessness, fatalistic 228
hepatitis C 116, 126–7
Heraclitus 217
herpes viruses 109, 151–2, 160, 174
hippocampus 99, 121–2, 162
histamine 15
HIV 44
Hobbit, The (Tolkien) 21
Hollow Men, The (Eliot) 170
homeostasis 33, 54, 56
Hooke, Robert 8–9, 17–18, 20
hormones 64, 82, 148, 152, 215–16; *see also specific*
      hormone
hygiene hypothesis 65–6
hypersensitivity 57, 64, 66, 109, 143, 147–8, 154
hypersomnia 135
hypothalamus 147–8
hysteria 98, 106

'I am, therefore I think' 49, 54
Icahn School of Medicine, Mount Sinai, NY,
      USA 120, 149

illusions 51–2, 165
immune privilege 19, 22
immune system; *see also* defence supersystem;
      inflammation/inflammatory response;
      sickness behaviour; *specific parts*
  adaptive 14, 16–17
  adjuvant activation of 149
  brain waste, analysing 27
  cancers, response to 19
  communication 12–13, 15, 16–17
  gut microbiome links 82
  hygiene hypothesis 65–6
  immune-brain loop 32–4
  immunoceptive inference 62–3
  inflammatory bias 65
  innate 14, 15–16
  memory 13, 14, 17
  modes of action 13
  nervous system, similarities to 17
  paper cut example of working 14–17
  phagocytosis 70
  PPRs 14, 15, 38, 40
  predictive processing 62–3
  self–non-self, distinguishing between 12
  as 'seventh sense' 48
  sleep impacts 214–15
  stress activation of *see* stress response
  threat detection 14–15, 17, 20, 37, 61, 65–6
  time in nature, impacts of 212
  twentieth-century discoveries 12–13, 14
immune tolerance 79, 83
immunoception 57–60
immunometabolic depression 135–6
immunopsychiatry 123
immunosuppressant drugs 61–2, 143
immunotherapy 104, 106, 115
Importance of Being Earnest, The (Wilde)
      156
indoleamine-2-3-dioxygenase (IDO) 121
infection bias 45–6
Inflamed Feeling, The (Lekander) 43
Inflamed Mind, The (Bullmore) 118
INFLAMED trial 134–5
inflammageing 179
inflammation/inflammatory response; *see also*
      sickness behaviour
  anger influencing 144–5
  and autoimmune diseases *see* autoimmune
      diseases
  birth method impacts on 83

brain, impact on *see* anti-NMDA receptor
  encephalitis; autoimmune psychosis;
  delirium; encephalitis, autoimmune;
  neuroinflammation; sickness behaviour
chronic *see* chronic inflammation
classical features of 16
defence system memory of 33–4, 142, 143–4,
  221
depression links *see* depression, clinical,
  inflammation links
detecting in others 42–3
dietary impacts 195, 196, 203
drug impacts 132–3; *see also*
  anti-inflammatories
emotional impacts 144–6
gratitude practice impacts 224
gut microbiome links 78–9, 86, 87, 89, 195, 196
importance in body's defence 187
inflammatory reflex 33
markers 118, 129–30, 131, 132, 133, 144
metabolism link 135–6
microglia action 30
movement impacts 208–10
neurogenesis, impact on 122
paper cut example of working 15, 22
reward processing, impacts on 126, 134, 153
sleep impacts 215
stress links 64–5, 111, 148–9, 152
time in nature, impacts of 225
walking, impacts on 42
inflammatory bias 65
inflammatory bowel disease 128
inflammatory reflex 33
infliximab 132
insomnia 135
insula 33–4, 125–6, 142
interferon 126–7
interferon-alpha (IFN-α) 116
interleukin-2 (IL-2) 144
interleukin-6 (IL-6) 40, 114–15, 120, 128,
  129–30, 132
interleukin-12 (IL-12) 128
interleukin-23 (IL-23) 128
International Schizophrenia Consortium 110
interoception 55; *see also* immunoception
intravenous immunoglobulin (IVIG) 101, 106–7
inulin 87
Israel Institute of Technology, Haifa, Israel
  33–4, 142
Iwasaki, Akiko 160–1, 163

Janeway, Charles 13–14
Jenner, Edward 11
Julius Caesar (Shakespeare) 93
Karolinska Institutet, Stockholm, Sweden
  35–6
kefir 196, 197
ketamine 100
Khan, Hazrat Inayat 139
Khandaker, Golam 129–30, 131
Kharkov Imperial University, Russia (now
  Ukraine) 70
kimchi 196, 197
King, Martin Luther 220
Kipnis, Jonathan 24–5, 26, 27–8, 47–8
Koch, Robert 12
kombucha 196, 197
Kraepelin, Emil 104
kynurenine pathway 121, 133, 163

Lactobacillus spp. 71, 74–5, 82, 198
Langerhans cells 15*n*
Langerhans, Paul 15*n*
latency 151–2
Law of Segregation 130–1
Lawson, Deodat 104–5
Leeuwenhoek, Antonie van 9
Lekander, Mats 35–6, 37–9, 41, 42, 43, 46–7
Lennox, Belinda 102, 103–4, 106–7
leprosy 44
Libby, Peter 178
light 216
light microscope 9
lipopolysaccharide (LPS) (endotoxin) 38–9, 43,
  122–3, 123–5
liver diseases 116, 179
Lives of a Cell, The (Thomas) 68
loneliness 154, 231–2
long Covid 157, 158–64
Longfellow, Henry Wadsworth 3
Louveau, Antoine 24–5
loving for defence supersystem
  accepting control and responsibility 228–9
  asking for help 221–2
  being flexible 227–8
  cultivating balance 229–30
  forming good habits 229–30
  gratitude practice 224
  growth mindset 228–9
  handling stress 226–8

loving for defence supersystem – *cont.*
  loving others 230–2
  loving our biosphere 232–3
  loving yourself 221–2
  mindfulness 225–6
  resilience 222–3
  social connections 230–2
  voicing problems 223–4
  writing feelings/thoughts down 223–4
lumbar puncture 26
lupus 65, 107–8, 114, 128
lymph nodes 16, 27, 113, 140, 148, 149
lymphatic vessels 16, 24–7
lymphocytes 209

Macbeth (Shakespeare) 35
McInnes, Iain 113–15, 128
Mackenzie, John Nolan 143
macrophages 12, 15, 16–17, 30, 31
Magendie, François 6
Maitland, Charles 11
major depressive disorder (MDD) *see*
  depression, clinical
malaria 10, 71
Man's Search for Meaning (Frankl) 145–6
mast cells 15, 22
mating 144
meditation 225
memory loss 159, 162; *see also* Alzheimer's
  disease; dementia
Mendel, Gregor 130
Mendelian randomization 130
meningeal immune system 23–8
meninges 23
mental health/illness; *see also specific illness*
  asking for help 221–2
  factors impacting
    birth, method of 83
    breast feeding 84
    diet 84, 200, 201, 222–3
    early life infections 64
    emotional regulation 145
    energy-generating molecules 84
    fitness and muscle strength 213
    movement 207–8, 223
    prefrontal cortex function/dysfunction 76
    self-efficacy 222
    sleep 223
    time in nature 212
  gut microbiome links 74–7, 84–6, 88–9

managing *see* diet for defence supersystem;
  loving for defence supersystem;
  movement for defence supersystem
  predictive processing view of 56–7
  stress links 64–5, 66, 153–4
  treating 201
mental resilience 84–5, 194, 202, 222–3, 227
metabolic diseases 135–6, 186–7
metaphors and analogies
  brain as scientist 53, 55
  car 180
  cup 111
  gut garden 201
  Ministry of Defence 63–4, 185–6
  Tom, Jerry and Terry 68–9
  triangle 168
  Unholy Trinity 89
  wheel 138, 186–7, 210
Metchnikoff, Élie 12, 69–71, 72, 87, 196
metformin 164
miasma theory 10, 12
microbes 9, 12, 68–9, 70–2, 77–8, 89, 188, 233; *see*
  also gut microbiome; microbiome;
  pathogens; probiotics
microbiome 69, 77, 78, 82–3; *see also* gut
  microbiome
microbiome transplants 76–7, 88, 177, 180
microglia 29–32, 120, 125, 132–3, 152, 162, 175–6,
  179
Micrographia (Hooke) 9, 17–18
Micropia museum, Amsterdam, Netherlands
  137
microscope, light 9
Miller, Andrew 132
mind-body dualism 18, 20, 32, 221
mind, location of the 5, 6, 18
mindfulness 225–6
minocycline 132–3
miso 197
mononucleosis, infectious 160
Montagu, Lady Mary Wortley 10–11
mood 37, 38, 115, 144–5, 194, 207–8, 213; *see also*
  depression, clinical
movement for defence supersystem
  benefiting all subsystems 206
  biological rhythms 217–18
  detriments of sedentary lifestyle 206–7
  exercise snacks 211–12
  guiding principles 211, 218–19
  how to move 211–12, 218–19

and inflammation 208–10
key takeaways 218–19
microbial motivation 204–6
and mindfulness 225–6
in natural surroundings 212, 219
play behaviour 213, 218, 219
positive impacts on brain function 207
positive impacts on gut microbiome 210
positive impacts on immune system 209
positive impacts on mental health 207–8
positive impacts on physical health 208–9
regular movement 218
and rest 213–17, 219
and threat detection 217–18
training 213
where to move 212, 219
WHO recommendations 211
Mucor fungus 9
multiple sclerosis (MS) 107, 161–2
Mycobacterium tuberculosis 12
myelin 76, 161

Najjar, Souhel 101, 103
National Health Service, Wales, UK 174
natto 197
natural killer cells 151, 209, 214–15
nature, time in 212, 219, 225
Nedergaard, Maiken 26
Nelmes, Sarah 11
nerves
  amygdala 75
  cranial 4, 75, 76
  fight-or-flight response 148
  in the gut 81
  gut microbiome interaction 205
  immune system interaction 22, 119, 125–6, 165
  myelination 76, 162
  shingles virus 151–2
  as signal highways 6, 7–8, 32–3, 64
  vagus 74–5, 81, 82
nervous system 6–8, 17, 36–7, 60–1, 76, 81–2,
    142–3, 148; *see also* defence supersystem;
    specific part
Netherlands Study of Depression and Anxiety
    (NESDA) 135
neurodegeneration 22, 87–8, 158, 162, 171, 177,
    180, 187; *see also* Alzheimer's disease;
    dementia
neurodevelopmental conditions 31–2; *see also*
    schizophrenia

neurodiversity 85
neurogenesis 121–2, 162
neuroinflammation 39, 79–80, 125, 132–3, 162–3,
    175, 176, 179; *see also* encephalitis,
    autoimmune
neurons; *see also* synapses
  anti-neuronal antibodies 102, 102n, 103–4,
    106–7; *see also* anti-NMDA receptor
    encephalitis
  CSF washing 26
  discovery of 6
  function 6, 28
  glial cell care of 29
  immune system interaction 33–4, 81–2,
    144
  inflammation *see* neuroinflammation
  myelination 161, 162
  neurodegeneration *see* Alzheimer's disease;
    dementia; neurodegeneration
  neurogenesis 121, 162
  as signal highways 7–8
  structure 6
  T-cell protection of 22
neuroplasticity 30–1, 84
neuropod cells 81
neuropsychiatric lupus 108
neurotransmitters 6, 82, 86, 120–1, 126, 144, 163,
    205
neutrophils 15, 149–50
Nixon, Richard 72
NMDA (N-methyl-D-aspartate acid)
    receptors 99–100, 167; *see also*
    anti-NMDA receptor encephalitis
nociceptors 7
Noma Guide to Fermentation, The (Redzepi and
    Zilber) 199
Noma restaurant, Copenhagen, Denmark
    198–9
non-alcoholic fatty liver disease 179
noradrenaline 148
norepinephrine 82

obesity 135–6, 187
olfaction 42–3, 158
'Olfaction and Prejudice' (Zakrzewska) 47
oligodendrocytes 29
oligosaccharides 83–4, 189
optimism, tragic 229
Osborne, Sarah 105
osteoporosis 179

pain
    chronic 157–8, 166–7
    pain reprocessing theory 166–7
    paper cut example 7–8
    placebo effect 144
    sensitivity to 22, 37, 61, 120, 126
Painful Truth, The (Lyman) 144
Parris, Betty 105
Pasteur Institute, Paris, France 69–70
Pasteur, Louis 12, 70
pathogens
    antigens 16
    behavioural impacts 68–9, 79–80
    body's defences against; *see also* immune
        system
        antibodies 17
        gut microbiome 78–9
        sickness behaviour 41–2
        stress response 149–50
    Koch's discoveries 151
    paper cut example of body's response to
        14–17
    present at birth 83
    spread of 12, 70–1, 71–2
    stress impacts 151
pattern-recognition receptors (PRRs) 14, 15,
    38, 40
Pavlov, Ivan 61
PCP ('angel dust') 100
Penicillium 72
Penninx, Brenda 134–6
Pepys, Samuel 9
perception 50–60, 66–7, 100, 165, 167; *see also*
    delusions; hallucinations
peri-arteriolar LEPR osteolectin cells 209n
periodontitis 174–5
peripheral nervous system (PNS) 7, 32
pets 212
phagocytosis 70
Phipps, James 11
phytochemicals 195
pia mater 23
pineal gland 18
placebo effect 143–4
plague, bubonic 10, 65
planetary health diet 195–6
plasma exchange 101
play behaviour 202, 213, 218, 219; *see also*
        movement for defence supersystem
Pollan, Michael 202

poo *see* faecal transplants
positivity, toxic 228
post-traumatic stress disorder (PTSD) 57, 153
post-viral illnesses 157, 158–64
poverty 154, 231–2
Prather, Aric 214
prebiotics 84, 87, 164, 190–6, 200, 202–3
prediction error 51, 52–4, 165, 167
predictive processing (coding) 50–5, 56–7, 61–3,
        66–7, 165–7, 207
prefrontal cortices 76
pro-inflammatories 116, 126–7
probiotics 71, 72, 73, 74–5, 84–5, 87, 164, 180,
        196–200, 203
Prolongation of Life, The (Metchnikoff) 71, 196
psoriasis 108, 115, 128, 179
psoriatic arthritis 115
psychedelic drugs 72
psychiatric diseases *see* mental health/illness;
        specific disease
psychobiotic diet 200
PsychoNeuroImmunology Research Society
        (PNIRS) 123
psychosis
    autoimmune psychosis 103–9, 112; *see also*
        anti-NMDA receptor encephalitis
    definition of 96
    genetic links 110
    gut microbiome link 109–10
    long-term *see* schizophrenia
    stress links 110–11
    toxoplasmosis link 79–80

quinolinic acid 121

rabies 69, 80
Radford-Smith, Daniel 84
Raggio, Samantha 93–7, 103, 107
Raison, Charles 132
Ramón y Cajal, Santiago 6, 29
randomized control trials (RCTs) 131–2
randomized trials 130–1
reactive oxygen species (ROS) 121
realism 228
redundancy 125
Redzepi, René 198–9
reflexes 7
resilience, mental 84–5, 194, 222–3, 227
rest 213–17
reward processing 126, 134, 153

rheumatoid arthritis 65, 114–15, 127, 128, 179
rheumatology 113–14
rhythm, circadian 216
rhythms, biological 217–18
rituximab 106–7
Rolls, Asya 142
Russo, Scott 120

Salem Witch Trials, MA, USA 104–6
Salmonella enterica 39–40
Sapolsky, Robert 226, 227
SARS-CoV-2 virus *see* Covid-19
sauerkraut 196
scarlet fever 71
Schafer, Dorothy 31
Schaller, Mark 45
schizophrenia; *see also* anti-NMDA receptor
        encephalitis; psychosis
    autoimmune conditions link 107–9
    cumulative risk 110
    definition of 96
    dysbiosis links 109–10
    first descriptions of 104
    genetic links 110
    infections links 108–9
    inflammation as common mechanism 109
    neuronal pruning links 31
    stress links 110–11
Schwartz, Michal 21–2
SCOBYs (Symbiotic Cultures of Bacteria and
        Yeast) 197
self-efficacy 154, 165–6, 213, 222, 226–7
Selye, Hans 147
senses 20, 36–7, 42–3, 47, 48, 50–3, 158–9, 212
'Serenity Prayer' 229
serotonin 82, 120–1, 163
Shakespeare, William 5, 35, 93
Shatz, Carla 31
shingles 151–2, 160, 174
short-chain fatty acids (SCFAs) 78, 82, 84,
        193
sickness behaviour
    author's experiences of 39–41, 123–5, 126
    brain's response to immune system 141
    depression links 122–3, 123–5, 126–7
    description 37, 40–1, 119, 214
    importance in body's defence 39
    Lekander's studies 37–9
    mode of action 119–20
    pathogen-avoidance 41–2

pathways involved 125
post-viral illnesses 161
preceding depression 119
SINAPPS2 trial 106–7
sirukumab 128, 132
skull channels 28
sleep 214–17, 219
smallpox 10–11, 71
smell, sense of 42–3, 158
social biasing 44–6
social engagement 38, 56, 76–7, 85, 86, 120, 154,
        230–2
socioeconomic status 154, 231–2
Soehnlein, Oliver 178
somatic nervous system 7
Sonnenburg, Justin and Erica 192–3
Sonnenburg Lab, Stanford University, CA,
        USA 192–3, 196, 200
sourdough bread 197
spinal cord 6, 7, 26
spinal taps 26, 98
spleen 33, 113, 140, 148
Standard American Diet (SAD) 191
Stanford University, CA, USA 31, 174, 192–3,
        196, 200, 226
stem cells 98, 209
steroids 99, 101
Stevens, Beth 31, 176
Stockholm University, Sweden 35–6
Strength to Love (King) 220
stress, psychological; *see also* stress response
    acute 148–50
    chronic 150–5, 226
    constancy of, modern-day 66, 141, 150–1,
        226
    defining 147
    eustress 227
    gut microbiome, impact on 74–5
    inflammation links 64
    mental illness links 64–5, 66, 153–4
    neurogenesis, impact on 122
    psychosis, causing 110–11
    reducing 226
    resilience to 84–5, 120, 194, 222–3
    reward response, impact on 153
    socioeconomic factors 154, 231–2
    taking control of 226–7
    voicing problems 223–4
    whole-body involvement 74
stress resilience 84–5, 120, 194, 222–3

stress response
  during acute stress 147–50
  birth method impacts 83
  during chronic stress 150–4, 163–4
  description 139–40
  fight-or-flight-or-freeze response 75, 140, 147–8
  gut microbiome impacts 73, 74
  mental illnesses, impacts on 153–4
  self-efficacy and outlook impacts 226–7
  whole-body involvement 140–1
stress-vulnerability model 111
strokes 178, 186
subarachnoid space 23, 25, 26
Summa Logicae (William of Ockham) 49
sunlight 216
synapses 6, 7, 29, 30–2, 99, 176; *see also* neurons
syphilis 44, 71

T-cells 16–17, 22, 24–5, 30, 149
tai chi 225–6
Taquet, Maxime 163
Task, The (Cowper) 188
taste 62
tau tangles 172–3, 178
taurine 86
tempeh 197
teratomas 98–9, 106
Theologian's Tale, A (Longfellow) 3
Thomas, Lewis 68
threat detection; *see also* sickness behaviour; stress response
  anger impacts on 144–5
  by brain 7–8, 17, 20, 28, 60–1, 127, 143, 144
  chronic inflammation impacts on 127–8
  chronic pain impacts on 166
  by defence supersystem 60–1, 63–4, 144, 147, 185–6
  hypersensitivity *see* hypersensitivity
  by immune system 14–15, 17, 20, 37, 61, 65–6
  microbiota impacts on 75, 78–9, 200
  modern-day barriers to 65–6
  play impacts on 213, 217–18
Through the Looking-Glass (Carroll) 204
thyroid disease, autoimmune 108
Tituba 105
Tolkien, J. R. R. 21
toll-like receptor 4 (TLR4) 38–9
toxic positivity 228
Toxoplasma gondii 68–9, 79–80

tragic optimism 229
trauma 57, 153–4
Treatise on Man (Descartes) 18
tryptophan 120–1
tuberculosis 12
tumour necrosis factor alpha (TNF-α) 114–15, 132, 179–80
tumours 19, 98–9, 106
typhoid fever vaccine 39–40
typhus 71, 145–6

Unholy Trinity metaphor 89
University College Cork, Ireland 73–4, 87–8, 177, 200
University Medical Centers (UMC), Amsterdam, Netherlands 134–5, 136–7
University of Bristol, England, UK 129–30
University of California, Los Angeles, USA 224
University of California San Diego School of Medicine, USA 192
University of California, San Francisco, USA 214
University of Cambridge, England, UK 133, 163, 178
University of Exeter, England, UK 212
University of Oxford Botanic Garden, England, UK 191
University of Oxford, England, UK 5, 84, 102, 163
University of Pennsylvania, Philadelphia, USA 98, 204–5
University of Pittsburgh, PA, USA 80
University of Rochester, NY, USA 26
University of Southampton, England, UK 174–5
University of Texas, USA 209
University of Tokyo, Japan 144
University of Toronto, Canada 175
University of Virginia, Charlottesville, USA 24
ustekinumab 128

vaccinations 11–12, 13–14, 37–8, 39–41, 148–9, 174
vagus nerve 33, 74–5, 81, 82
varicella-zoster virus 151–2, 160, 174
variolation 10–11
vascular dementia 178
Vibrio cholera 12
Vincent, Angela 102
Virchow, Rudolph 29

viruses 151–2, 214–15; *see also* pathogens;
    post-viral illnesses; *specific virus*
vision 52–3
vitamins 195
voicing problems 223–4

Walker, Matthew 214, 215, 216
Washington Post 107–8
Weaver, Donald 175
Why We Sleep (Walker) 214
Wiezmann Institute, Israel 21–2
Wilde, Oscar 156
William of Ockham 49
Williams, Abigail 104–5

Willis, Thomas 4–6, 9, 17, 20, 43
witchcraft, fear of 105–6
wonder and awe 89–90, 225
World Health Organization (WHO) 211
Wren, Christopher 4, 6, 9
writing feelings/thoughts down 223–4

Yale University, New Haven, CT, USA 13,
    160–1, 163
Yersinia pestis 10, 65
yoghurt 70–1, 72, 87, 196, 197

Zakrzewska, Marta 47
Zilber, David 197–200

About the Author

Dr. Monty Lyman is a medical doctor, researcher and author who specializes in the relationship between the mind and immune system. He is an Academic Clinical Fellow at the University of Oxford.

His first book, *The Remarkable Life of the Skin*, was shortlisted for the Royal Society Science Book Prize, was one of the *Sunday Times* 'Best books of 2019' and was a Radio 4 'Book of the Week'. An essay from his second book, *The Painful Truth*, won the 2020 Royal Society of Medicine's Pain Medicine Prize.